Engineering Design Principles

D0182794

Engineering Design Principles

Kenneth S. Hurst

University of Hull

A member of the Hodder Headline Group
LONDON • SYDNEY • AUCKLAND

Copublished in North, Central and South America by
John Wiley & Sons Inc., New York • Toronto

First published in Great Britain in 1999 by
Arnold, a member of the Hodder Headline Group,
338 Euston Road, London NW1 3BH

http://www.arnoldpublishers.com

Copublished in North, Central and South America by
John Wiley & Sons Inc.,
605 Third Avenue,
New York, NY 10158-0012

British Library Cataloguing in Publication Data
A catalogue record for this book is available from the British Library

Library of Congress Cataloging-in-Publication Data
A catalog record for this book is available from the Library of Congress

ISBN 0 340 59829 8
ISBN 0 470 23594 2

1 2 3 4 5 6 7 8 9 10

Commissioning Editor: Matthew Flynn
Production Editor: Wendy Rooke
Production Controller: Sarah Kett
Cover Design: Mouse Mat

Typeset in 10/12 pt Times by Phoenix Photosetting, Chatham, Kent
Printed and bound in Great Britain by J W Arrowsmith Ltd, Bristol

What do you think about this book? Or any other Arnold title?
Please send your comments to feedback.arnold@hodder.co.uk

Contents

1 Introduction to engineering design

In this introductory chapter the engineering design process which is covered in detail later is defined. A historical perspective is taken to explain the need for a formal process and the complexity of current engineering is outlined. A definition is given for both the engineering design process and the duties of an engineering designer. Design is defined as a technology, not a science, and accepted models of the process are presented. Finally the levels of communication necessary for successful engineering design are illustrated.

1.1 Historical perspective

The study of history is often very illuminating and two main purposes are served. The mistakes made by previous generations should not be repeated. Also, a vast body of knowledge has been gathered over the centuries, which can be usefully employed in the present day. There are important lessons here for engineering designers, since most new products are not completely new inventions but are new applications and combinations of existing technologies. A study of the history of science and technology serves to enhance our understanding of modern day engineering and helps to prevent the all too real temptation of 're-inventing the wheel'.

It is surprising how far back in time this study can usefully delve. As an example consider the invention of the force pump, the earliest description of which was given by Philo of Byzantium in the second century BC. Figure 1.1 is a reproduction of Philo's drawing. There is little evidence of refinement or aesthetic considerations but all the essential principles are presented and the design is unexpectedly complex. Water flows into the partial vacuum created by the upward motion of the piston and on the down stroke, with the valves reversed, the water is forced up the pipe into the tank. When an invention such as the force pump first surfaces it is considered to be a dynamic product. That is to say that conceptually it is a significant advance or step change from anything which has gone before. Also, there remains scope for significant product development in a dynamic product which is not possible in those products defined as static.

As is the case with most useful inventions, the force pump was subsequently refined and next appeared in the form illustrated in Fig. 1.2, which is Hero's force pump from the first century AD. At this stage of development the pump would be considered to be conceptually static since the later design follows the previous design. The refinements which are most noticeable include the replacement of the two pipes for conveying water to the tank with one, the single actuation beam pivoting in the centre and the introduction of a nozzle. The nozzle was introduced specifically for fire fighting applications although it was many centuries later that the pump was mounted on a chassis in order to travel to the scene of a fire. Thus the conceptually static force pump could again be considered as dynamic when the idea for mobility was first suggested, many hundreds of years after the initial design.

Figure 1.1 Philo's force pump (Reproduced from Carra de Vaux, *Les pneumatics de Philon*, p. 217)

To prevent the wrong impression being created by this example it is important to realize that not all design work is innovative in nature and many product developments are incremental. In fact the majority of an engineering designer's life is spent making relatively minor improvements to existing products.

In those early days, artisans such as Hero and Philo conceived ideas almost completely in their minds and generally worked in isolation before communicating the finished concept to others. Thoughts now, as then, can be verbal but are more often than not visual and three-dimensional, particularly for engineers. Unlike Hero and Philo the modern design engineer must be able to express thoughts clearly and communicate them constantly, throughout the whole design and development process, both within the design team and outside. This communication process inevitably involves a great deal of sketching and some skill in this area is thus essential for a designer. It is very important therefore that the student engineer develops by practice the skill of sketching quickly in 3D.

The level of scientific and technical knowledge possessed by Hero and Philo was limited when compared with today's understanding. Nevertheless, study of the two figures

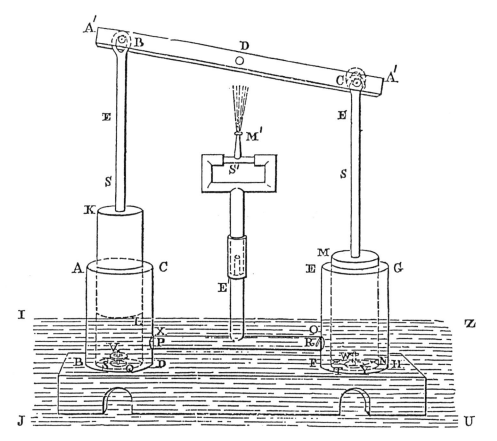

Figure 1.2 Hero's force pump with adjustable nozzle (Reproduced from a facsimile of the 1851 Woodcroft edition, introduced by Marie Boas Hall, London 1971, drawing of 'The Fire-Engine', p. 44)

showing the force pump indicates that they must have had an intimate knowledge of materials, the engineering sciences and manufacturing processes which were current at the time. These artisans did not openly employ a formal engineering design process. However, with the considerable advances made in materials and manufacturing technology, increased knowledge in engineering science, ever more stringent environmental consider- ations, increasing competition, greater emphasis on energy efficiency and increasing sophistication required of today's products, a formal engineering design process has become essential. They must also have had at least an appreciation of economic constraints since people with buckets could probably perform the same tasks as a force pump for a much lower initial investment!

It should be said at this point that it comes as a considerable shock to most young engineers when they first realize that only a very small percentage of decisions made by a design engineer are based on complete knowledge of the engineering sciences. The knowledge used by a design engineer is extremely broad and varied in nature. It is true that part is derived from science but a great deal comes from testing and evaluation and on observations of materials and systems.

In the days of Hero and Philo engineering effort made a significant impact on people's lives relatively infrequently and so was regarded as marvellous. Also, the practitioners are remembered. This is not very often the case today, even though engineers have a disproportionate effect on the kind of world we live in. Less than 1% of the population are engineers and yet virtually everything we see if we look around is man-made and has been designed to be that way.

1.2 Engineering design definition

In the study of science we seek to develop theories that explain natural phenomena. Scientific theories consist of a statement or set of statements that define some kind of ideal or theoretical system. These scientific principles, which are self-evident in the natural sciences, are also employed in the engineering sciences. Engineering science subjects such as thermodynamics, mechanics and materials science are generally based on established scientific principles like the first and second laws of thermodynamics, Newton's laws, and atomic and molecular theories of matter respectively.

Engineering design is quite different since theories and hypotheses cannot be developed or tested by laboratory experiments. Engineering design involves much broader issues including the consideration of people and organizations. It must therefore be regarded as a technology. This is particularly so since no single absolute answer can be found for any problem which involves both design decisions and compromise, since almost inevitably design parameters are contradictory.

Having established that engineering design is a technology it is necessary to present a definition. Many attempts at a definition have been made, particularly in the search for a snappy, short definition, but all attempts to date have been defeated. The dictionary definition of design is often 'to fashion after a plan', which tells us very little about the way of working that we call engineering design. What follows is an amalgam of definitions for both the process and practitioners taken from the UK based Institution of Engineering Designers and the engineering design lecturer organization, SEED Ltd (Sharing Experience in Engineering Design).

> Engineering design is the total activity necessary to establish and define solutions to problems not solved before, or new solutions to problems which have previously been solved in a different way. The engineering designer uses intellectual ability to apply scientific knowledge and ensures the product satisfies an agreed market need and product design specification whilst permitting manufacture by the optimum method. The design activity is not complete until the resulting product is in use providing an acceptable level of performance and with clearly identified methods of disposal.

In order to increase our understanding of design it is helpful to extend this definition and to identify and highlight the main characteristics of engineering design:

- Trans-disciplinary
- Highly complex
- Iterative.

Most engineering design is now a trans-disciplinary team effort and the distinctions between the traditional disciplines, mechanical, electrical, electronic, civil and even chemical engineers are becoming blurred. Relatively new areas of engineering specialization, such as control and software engineering should be added to this list.

Consider for example automobiles, which not so very long ago were the sole province of mechanical engineers. Complex engine management systems, anti-lock braking systems, active suspension systems, four-wheel steering, air bags, and automatic seat belt tensioning are just some of the new developments. These systems are highly complex and require input from many different kinds of engineers for their optimum design. Also, the selection of the appropriate technology for each part of a truly integrated design has become critical to the success of the product. Only engineers with a broad understanding of all potentially useful technologies and all the issues involved can make optimal decisions.

As an illustrative example we can usefully consider anti-lock braking systems (ABS). Perhaps because traditionally automobile design was the province of the mechanical engineer, the first anti-lock braking systems introduced were purely mechanical in nature. Although the performance of mechanical units was considered to be adequate in the early stages, their performance has since been surpassed by integrated systems which include software, electronic and mechanical technologies. Purely mechanical systems cannot match the performance levels which can be achieved by integrated designs.

An ABS, as the name suggests, improves braking performance by preventing wheel lock. A modern system can be seen in Fig. 1.3 with (a) being the general layout and (b) being the electronic circuit diagram. In the complete description of the system a hydraulic circuit diagram would also be required along with component details. Such systems allow braking to occur without impairing directional control and shorten braking distances considerably. This is accomplished by sensing the speed of each wheel along with wheel acceleration, comparing this to forward speed and modulating brake pressure accordingly.

The complexity of the complete system shown in Fig. 1.3 (a) and (b) serves to illustrate the need for inter-disciplinary engineering design teams. The main components in the general layout are:

(1) Front wheel sensor
(2) Front pulse wheel
(3) Hydraulic modulator
(4) Control unit
(5) Rear wheel sensor
(6) Rear pulse wheel
(7) Indicator lamp
(8) Brake tubes.

The induction sensors are used to signal wheel speed information to the control unit (computer). Signals are received by the control unit as sinusoid voltages and converted to digital signals for processing in the logic circuits. The main components in the electrical layout are:

 (1) Battery
 (2) Ignition switch
(10) Alternator

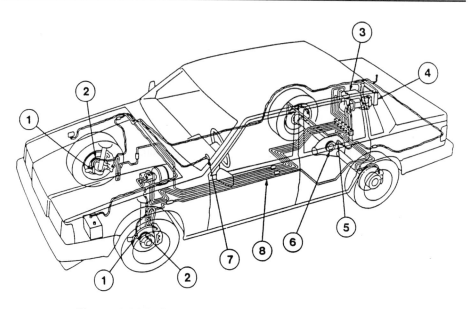

Figure 1.3(a) General layout of a Volvo ABS system

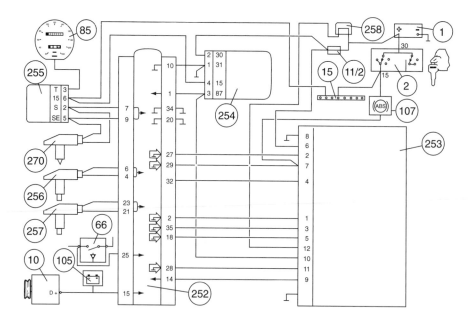

Figure 1.3(b) Electronic circuit diagram for a Volvo ABS system

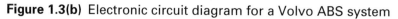

(11/2) Fuse
(15)   Distribution coil
(66)   Brake switch
(85)   Speedometer
(105) Charging lamp
(107) Indicator lamp
(252) Control unit
(253) Hydraulic modification
(254) Surge protection unit
(255) Speedometer converter unit
(256) Sensor – left front
(257) Sensor – right front
(258) ABS Fuse box
(270) Sensor – rear wheels.

It is not possible to give a complete and detailed account of the design of anti-lock braking systems in this text, nor is it desirable. However, it is important to note that a design such as this is very soon superseded. The detailed information presented regarding ABS is reproduced with the kind permission of Volvo Car Corporation, whose designs have become more sophisticated. For example, the S80 is available with the active chassis system DSTC (Dynamic Stability and Traction Control). This system uses a number of sensors, including a yaw angle sensor, to compare the way the car is handling with the way it ought to be behaving. DSTC then retards the appropriate wheel or wheels in order to stabilize the car. Volvo describe DSTC as an invisible hand which keeps the car on the road, even in extremely slippery conditions!

The purpose in using ABS as an example is purely to reinforce the stated definition of engineering design and to illustrate the technical and human interface complexities which are encountered in modern systems design. Along with this definition of the design process it is illuminating to consider the job description of an engineering designer. Although this can vary in detail, in general an engineering designer must be capable of dealing with the following:

- The production of practical design solutions starting from a limited definition of requirements taking into account many factors.
- The production of design schemes, analysis, manufacturing drawings and related documentation within defined timescales.
- The assessment of the design requirements of a particular component, system, assembly or installation in consultation with other departments.
- The production of designs which will favourably influence the cost and functional quality of the product and improve profitability and/or the company's reputation with customers.
- The undertaking of feasibility studies for future projects.
- Negotiations with vendors on aspects of bought out components and equipment, and with subcontractors or partner firms on interfaces.
- The assessment of the work of others.

The personal characteristics, derived from these responsibilities, which a design engineer must possess are:

- ability to identify problems
- ability to simplify problems
- creative skills
- sound technical knowledge
- sense of urgency
- analytical skills
- sound judgement
- decisiveness
- open mindedness
- ability to communicate
- negotiating skills
- supervisory skills.

These abilities and skills are possessed by everyone to a greater or lesser extent. They are developed in engineering designers over a period of time, mainly by the practice of engineering design and by exposure to the design process.

1.3 The engineering design process

The cost of a product, particularly in international markets, is only one factor which has a bearing on success. Reliability, fitness for purpose, delivery, ease of maintenance and many other factors have a significant influence and many of these are determined by design. Good design is therefore critical for success both in national and export markets and can only be ensured by adherence to a formal design process.

The engineering design process in its simplest form is a general problem solving process which can be applied to any number of classes of problem, not just engineering design. It must be remembered that the design process as outlined will not produce any design solutions. The aim in recommending a design process to adopt is to support the designer by providing a framework or methodology. Without such a process there is the very real danger that when faced with a design problem and a blank sheet of paper the young engineer will not know how to begin. The rigorous adherence to the process as outlined later will free the mind, which can become extremely cluttered during a project, so that more inventive and better reasoned solutions emerge.

A systematic approach permits a clear and logical record of the development of a design. This is useful if the product undergoes development and redesign. Also, the disturbing trend of law suits against companies and individuals often means that the designer must be able to prove that best practices were employed. This can best be established by reference to comprehensive supporting documentation, such as records of the decisions made and reasons why they were made.

If we accept the need for a systematic approach, how and in what order should we consider the influencing factors? There are several suggested systems which vary in detail but are basically similar. Figure 1.4 illustrates the design process as presented by Pahl and Beitz (G. Pahl and W. Beitz (1984) *Engineering Design*. London, Design Council) and Fig. 1.5 is that recommended by SEED. Study of the two figures reveals an underlying similarity with the basic process being to identify the problem, generate potential solutions, select from the solutions, refine and analyse the selected concept, carry out detail design and produce product descriptions which will enable manufacture. Quite

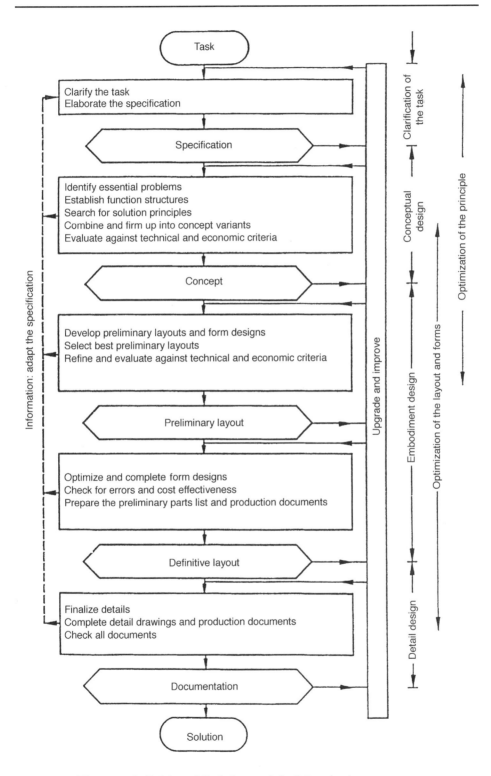

Figure 1.4 Pahl and Beitz's model of the design process

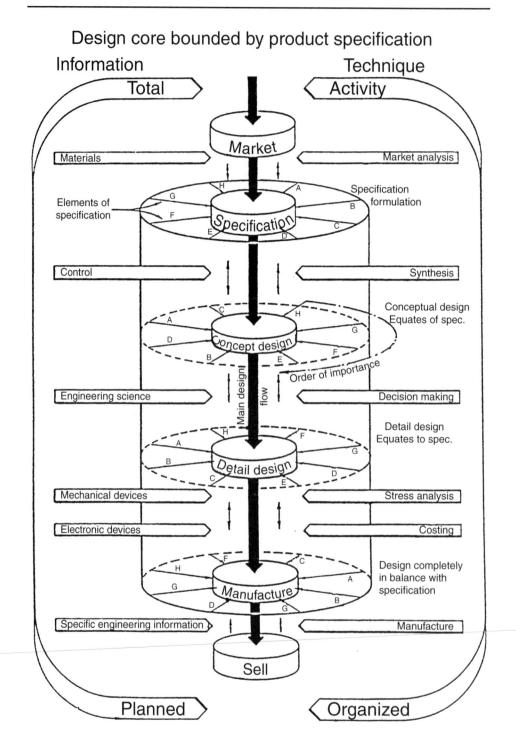

Figure 1.5 Pugh's model of the design process

obviously for both models to be complete they must be extended to include use and recycling or disposal.

The SEED (Pugh) model is the one we will be following throughout the text. As indicated by the return lines, design is an iterative process involving much back tracking and parallel activity. This is normal. The principle of iteration is the fundamental principle of the design process. Designing something new is like a voyage of discovery. As the design progresses, more and more information is discovered and more knowledge gained. If the designer does not iterate the new information, concepts emerging would not be acted upon. The systematic approach is not a series of instructions to be followed blindly. There is never a unique solution.

Some words of caution. Engineering design is not always a sequential process, nor can it be neatly divided into discrete activities each of which must be fully completed before the next is begun. This is why feedback loops are always included on any diagram of the design process. However, the reader should be aware that even this does not do justice to the necessary continual iteration and that all steps in the design process are often going on simultaneously. Also, an engineering designer is rarely completely satisfied with the solution arrived at. This is partly due to the principle of time. If a company is to maximize its profits from the labours of a design team then the shortest possible time must be taken in getting the product launched. It is inevitable therefore that with a second look the product could be improved. This lack of perfection often causes dissatisfaction and must be accepted as a consequence of working as an engineering designer.

The first and most important stage in the design process as outlined in Fig. 1.5 is the formulation of a Product Design Specification (PDS). This is especially important as international trade becomes simpler and competitiveness becomes harder to achieve. Companies must use a logical and comprehensive approach to design if they are to profit from their labours. Therefore an all encompassing problem definition which is used to audit and guide the remainder of the design process is essential.

The process of design is always the same and is not dependent on the size or complexity of the problem. However, it is almost always subject to unforeseen complications and a flexible design management approach is essential.

1.4 General example

As a simplified illustration of the design process consider the problem of building an extension to a court house. The brief stated that the whole of the works must be carried out with the existing court in full operation and, due to national terrorist activity, that bomb blast protection be provided. The requirement was:

- A new courtroom
- Offices, stores and amenities for ushers and clerks
- A new boiler room.

Specification   In order to develop a full and detailed specification of the problem many initial investigations were carried out. Drawings of the existing building were obtained but, as is normal, no original strength calculations were available and depth of foundation piles was not known. A site survey was undertaken and a geotechnical investigation revealed that the ground was poor down to depth of 21 m. Samples extracted from bore holes were laboratory tested enabling moisture content, chemical composition and particle sizes to be obtained.

Concept generation Having defined the specification, including the relevant British Standards for foundations (BS 8004), structural use of concrete (BS 8110) and notes on blast resilient designs, the next stage was to consider alternative concepts. After initial brainstorming and the consideration of many concepts only three were considered worthy of investigation. These were:

(1) add an additional storey;
(2) infill at ground floor level and extend at ground floor level;
(3) an additional two storey building linked with corridors.

Concept selection Concept 1 was ruled out since on investigation the existing piles were found to be fully loaded and the cost of strengthening the structure would be prohibitive. Concept 3 was ruled out because it was impossible to allow for sufficient daylight reaching existing windows. Concept 2 was selected because it presented the optimum solution, when compared with the specification, including customer requirements.

Detail design Having made this overall decision, further detailed investigation was necessary, with much engineering science and materials knowledge being employed. Decisions such as whether to use driven or augered piles and use ground beams or slabs had to be made. Above foundation level masonry stability, vertical, shear and bending loading capacities needed to be calculated.

Manufacture Once this detail design stage had been completed the construction phase could begin. This phase can be likened to prototype manufacture prior to mass production of products.

The project reported was very complex and was completed slightly ahead of schedule within the customer's budget. Many engineering projects do not go so well which, in most cases, can be traced back to a lack of adherence to the iterative step-by-step approach being advocated. In this case the design process was followed rigidly. Effective lines of communication established at the outset of the project were just as instrumental in the success of this project as the engineering expertise employed.

1.5 Engineering design interfaces

As outlined earlier it is essential that a design engineer has good communication skills. This can be further reinforced by study of Fig. 1.6 which gives an indication of the most frequently used lines of communication, both within the design department and outside. As explained earlier the design process begins with a design brief or Product Design Specification. This then is the major trigger which causes the design department to act.

Two broad types of communication can be identified, internal and external. Those internal communications with the design department may include defining input parameters for computation, discussion and information transfer with other relevant design groups, informing the drawing office by such means as scheme drawings and materials

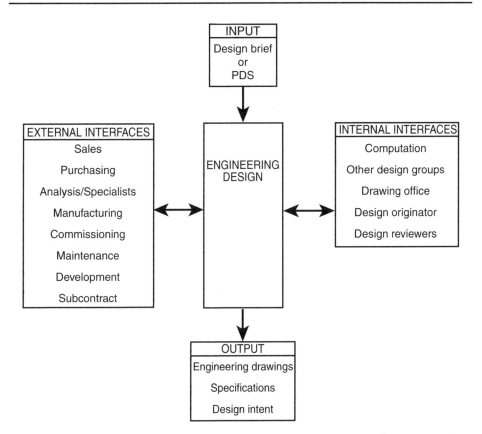

Figure 1.6 Company-wide engineering design interfaces

specifications, gaining approval for proposals from the originator and answering the questions of reviewers at design audit meetings.

External communications both with other departments and outside the company are easier to define than internal communications and are probably of greater importance. Those identified in Fig. 1.6 are the main lines of communication although many others exist. In detail the types of communication with other departments and outside are:

Sales   There is continuous two way communication between the design and sales departments. The sales department supply customer requirements and the design department supplies technical descriptions, performance data and predictions.

Purchasing   This is generally one way communication with the design department supplying the technical information essential for the purchasing department to buy in components.

Analysis/Specialists   Within companies there are many specialists who are often consulted by the design team. These may include standards, materials and stress analysis experts, amongst many others.

Manufacturing   Although this is indicated as a one way communication process in Figs 1.4 and 1.5, with design supplying working drawings to manufacturing, many other links exist. During all design review meetings at least one representative of the manufacturing department will be present to ensure optimum manufacturing methods are being specified by the design team. This is part of what has become known as concurrent engineering, which serves to shorten the time taken from initial concept to the production of the first products for sale. Also, manufacturing departments provide feedback to design and request design changes which ease manufacture.

Commissioning and Maintenance   This is generally a one way process with information being fed back to the design department when problems are encountered.

Development   In smaller companies design and development form one department which is an indication of how closely linked the two departments are. In the development department tests are carried out on particular aspects of design concepts generally by manufacturing the design and performing accelerated tests or by simulations. The results of these tests are fed back to the design team.

Subcontract   There are very few companies which have the facilities to manufacture fully everything they sell. Also, it is often cheaper, due mainly to economies of scale, for components to be bought in. It is necessary for the design team, in conjunction with the purchasing department, to communicate with potential subcontractors and to use their expertise.

    As already discussed, not only do design engineers need to communicate with many different people in many different departments they also require inter-disciplinary skills encompassing the many different fields of engineering. However, the required breadth of knowledge cannot be gained in the classical academic way. As illustrated in Fig. 1.7 industry requires engineers who have complete knowledge in all disciplines. By contrast traditional academic courses tend towards producing people who know everything about very narrow subject areas. There is also a very real danger that engineering designers will not develop the required level of detailed knowledge to complement the required breadth. Thus, in the final illustration in Fig. 1.7 the design engineer is shown as having broad knowledge complemented by 'ears' of detailed knowledge.

1.6 Principles

Throughout the book engineering design principles which are identified are presented at the end of each chapter.

Introductory principles

Iteration   Progress towards a solution should involve all the stages identified in order, but much backtracking is essential. This is the nature of engineering design.

Compromise   A perfect or single solution rarely emerges and the best that can be achieved is an optimum solution. That is a design which best satisfies the customer.

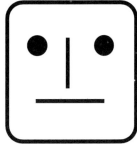

Classical aim of academic education   What industry wants

Danger – no detail knowledge   Aim – design engineer

Figure 1.7 Graduate profiles

Complexity   Engineering is a technology, not a science, so along with the engineering science knowledge used the importance of communication, teamwork, project management and ergonomics cannot be underestimated.

Responsibility   There is the potential for many failures to occur due to negligence or oversight and the ultimate responsibility for safe and correct 'products' rests squarely on the shoulders of the professional engineering designer.

Simplification   In general the simplest solution is the best and all professional engineers seek elegant and simple solutions.

As a final thought consider the alternative, humorous though cynical, design process suggested by Dr Glockenspiel:

(1) Euphoria
(2) Disenchantment
(3) Search for the guilty
(4) Punishment of the innocent
(5) Distinction for the uninvolved.

2 Problem identification

Here the processes necessary for the definition of a Product Design Specification (PDS) are detailed. As a prelude to writing the PDS much research must be carried out and much information gathered. This is a continual process and is described in Chapter 9. In this chapter the required contents of a PDS are explained and the format of a PDS illustrated by example. The writing of a PDS is the essential first step in every design project.

2.1 Introduction

If you were asked to design a corkscrew could you do it? Reference to Fig. 2.1, which illustrates many different types of corkscrew, probably convinces us that the answer to the question is yes. However, why are there so many fundamentally different types? How is it possible for different design teams to set out to design a corkscrew and end up with completely different devices?

In a little more detail the corkscrews illustrated in Fig. 2.1 are the plain corkscrew, with from left to right a double helix, lazy-tongs, the waiter's friend, a lever system and a screw pull. The double helix uses both left and right hand screws. One is inserted in the cork and the other forces the corkscrew against the neck of the bottle and removes the cork. The lazy-tongs illustrated provide a 4:1 mechanical advantage. Once the screw is inserted in the

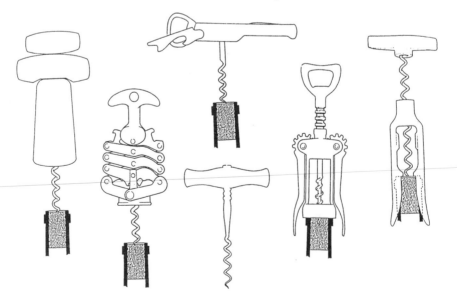

Figure 2.1 Corkscrews

cork the handle is pulled and travels four times the distance that the cork travels thus reducing the force required. The waiter's friend provides a mechanical leverage which is dependent on the length of the handle. When the screw is inserted in the lever system the levers rise. As they are pressed down the cork is extracted by pushing against the neck of the bottle. In the final device the screw is simply inserted in the cork and the turning continued. The cork 'climbs' the screw.

All of the devices described rely on the screw to be inserted in the cork. They differ in the mechanical advantage provided, in appearance, in complexity and in production cost. In order that a satisfactory product is designed the market need must be thoroughly researched and a technical specification reflecting customer requirements developed. In the case of a corkscrew constraints such as the mechanical advantage required, the appearance and the production (ex-works) cost must be specified.

In the solution of any design problem the design process begins with the defining of the boundaries within which a solution must be found. The project brief as presented to the design team is often incomplete. Hence, research often needs to be conducted and information sought before a full Product Design Specification (PDS) can be produced. Even if a full PDS is provided it is the duty of the designer to question the validity of that PDS.

This questioning approach can often make a customer alter their requirements. As an example consider the problem which was set as one of designing a corkscrew. If the original problem statement had been to design a device for removing a cork from a bottle then many more solutions are possible. Figure 2.2 illustrates two devices for removing corks which do not use a screw, the wiggle and twist extractor and an air pump. In application the two prongs of the wiggle and twist extractor are inserted between the walls of the bottle and the cork. By careful combination of pulling and twisting the cork is removed. The air pump employs a hollow needle which is pushed through the cork. Subsequent pumping action increases the pressure behind the cork and the cork is pushed out. Both of these valid devices were ruled out by the thoughtless problem statement which dictated that a screw be used.

As a final thought on this problem it is interesting to consider the problem statement as to remove wine from a bottle. More importantly, the new problem statement is as intended from the outset. If this is the intention then removing the cork may only be one category of solution! As further emphasis of the importance of a clear problem statement consider two wonderful engineering achievements.

The two photographs, Figs 2.3 and 2.4 show what was until April 1998 the longest single span suspension bridge in the world, the Humber bridge, and Concorde (the only supersonic airliner in the world) respectively. Both are elegant and simple in form and each is a

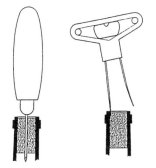

Figure 2.2 Cork extractors

Figure 2.3 Humber Bridge (Reproduced by kind permission of
The Humber Bridge Board)

Figure 2.4 British Airways Concorde, the flagship of the world's civil
aviation fleet (Reproduced with kind permission of British Airways)

magnificent feat of engineering which must be seen to be believed. However, neither has made a profit for their owners!

A first draft of a PDS must be developed before any attempt is made at generating solutions to a problem. This is an important discipline since so much time, effort and money can be wasted by providing a solution to the wrong problem.

Whilst it is desirable that a fully defined PDS be written before the design process starts it must be recognized that for many projects this proves impossible. The design process is iterative and the PDS must be regarded as a fluid document which will develop along with the design. This is indicated in Fig. 1.5 by means of the return arrows. The PDS is questioned at all stages and reference made to the customer as and when changes are suggested by the design team. However, the aim at the outset is to define the PDS as fully as possible.

It is extremely important that prospective customers are identified and that the language used in the PDS can be readily understood. Even within engineering each discipline, mechanical, electrical, electronic, civil and chemical, has evolved a specialist code not readily understood by other engineers. The customer may be involved in a totally different profession and yet must be able to understand fully the PDS.

It is the duty of the design team to verify that every function and constraint specified is relevant, correct and realistic. Consequently, it is essential that a thorough investigation of the problem is made by the designer before a solution is sought. For large, complex and diverse problems it is generally worthwhile breaking the project down into smaller, more manageable, sections.

In general there are two main tasks which have to be completed if a thorough identification of the problem is to be achieved:

(1) definition of the problem area;
(2) formulation of the exact problem.

The exact formulation of the problem involves the writing of a comprehensive PDS defining all the required *functions* which the solution must provide and all the *constraints* within which the solution must work. The information necessary for addressing these two tasks may be known or could be determined by calculation, by testing and by information search. Wherever possible a questioning approach should be employed and questions should be phrased in such a manner that a specific or numerate response is demanded. The information gathering process, which is a continual process, is explained in Chapter 9 and illustrated in Fig. 9.1. The information inputs required for the PDS are illustrated in Fig. 9.2.

2.2 PDS criteria

The main headings and criteria listed here and illustrated in Fig. 2.5 are intended to assist in the writing of the PDS. They are not to be regarded as an all embracing check-list which if followed blindly will completely define any PDS. Design projects are by their nature diverse and substantially different criteria are required from one project to the next. Nevertheless, the check-list will provide a good foundation upon which you, the student engineer can build. Once the project is begun you will find that many of the important criteria will suggest themselves. However, it is true that there is no substitute for experience and you should always be prepared, at any stage of the design process, to ask for help and guidance from experts such as component suppliers.

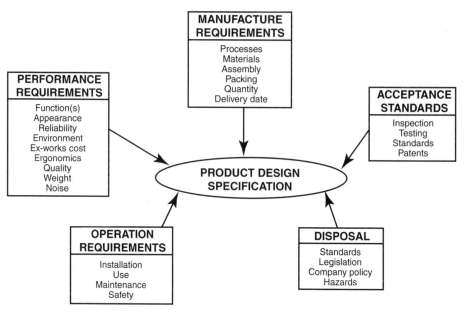

Figure 2.5 PDS criteria

The five main headings on Fig. 2.5, performance requirements, manufacture requirements, acceptance standards, disposal and operation requirements are now considered in detail.

Performance requirements

Function(s) There may only be a single main function which is to be provided by the product to be designed but this is unusual. More often than not multiple functions can be identified which can be divided into primary and secondary functions. These can vary in nature from mechanical, electrical, optical, thermal, magnetic and acoustic functions to name but a few. The primary function of an engine in a vehicle is to drive the wheels. Secondary functions, such as providing heating inside the vehicle and supporting alternators must also be listed.

Loading Loading can be divided into primary and consequential loading. Primary loads are due directly to the required function being provided. Shocks and vibration are generally consequent on the situation in which the product is used. Consequential loading is often very difficult to quantify without empirical data. Specified performance requirements should generally be met comfortably, with some performance to spare.

Aesthetics In some instances this is not important, particularly where the device or structure is not seen. However, for many consumer products or structures a pleasing elegant design is required and colour, shape, form and texture should be specified. All visible aspects must be in accordance with the nature of the product and reflect the corporate image of the company. Any statement in a specification which relates to the way a product will look is inevitably more qualitative than quantitative and should include analogy to qualities found in existing products or natural objects. It is possible to use techniques like golden section, which indicates that for aesthetic beauty any shape should be divided into two thirds and one third.

Reliability   The required design life, taking due account of routine maintenance, must be specified. This is usually done by specifying the number of operating cycles rather than in units of time. Within this number of cycles an acceptable level (%) of random failures or breakdowns is also specified. Where high levels of life expectancy of components exist and it is known that those components will be employed in a controlled environment, such as in electronic circuits, it is common practice to specify the MTTF (Mean Time To Failure) and the MTBF (Mean Time Between Failures). Where reliability is critical, redundancy, either active or stand-by, should be specified. Reliability is inextricably linked with maintenance, even if a maintenance free product is envisaged.

Environmental conditions   These include the temperature range, humidity range, pressure range, magnetic and chemical environmental conditions to which the product will be exposed. It is important to consider manufacture, store and transport environmental conditions along with the more obvious operating conditions. Also, any physical size restrictions should be specified. This is mainly dictated by the area available to the product when working but is often determined by considering transport and erection. The simplest form of expression for this constraint can be a diagram which forms an integral part of the PDS.

Ex-works cost   Companies sell products for the maximum price the market will stand which often bears little relation to the cost of producing that product. Hence, the maximum cost specified in the PDS and which the design team must work to, should be the production (ex-works) cost and not the selling price.

Ergonomics (Human factors)   If a product is intended for human use then account must be taken of the characteristics of those users. The design of the product and the tasks required of the product and the users must reflect their respective capabilities. The person/product interface, as identified in Fig. 2.6, must be carefully specified. Decisions are based on those functions which can be carried out by products and will vary as capabilities of machines increase. The functions carried out by the user are generally to sense a display, interpret it and make a decision and perform a controlling action.

The environment in which the product is to be operated should be specified carefully. For example, if noise levels are high then audible signals to which a user must respond may not be heard. Anthropometrics is the branch of ergonomics which deals with body measurements and it is normal to specify a user population who fall between the 5th and 95th percentile sizes in any particular respect. Any controls must operate in a logical or expected manner. Controls should be placed in easy reach of the operator.

Quality   The quality of the product should meet market requirements and the quality of all components should be consistent. All workmanship must be in accordance with the best commercial practices. Robust design practices should be used where possible. All materials and components shall be new and free from defects.

Weight   In some industries, such as aerospace, this is the most critical constraint. However, this is not always the case and weight is not always required to be a minimum. Generally in any product involving motion reduced weight is an advantage whereas a product where stability is critical may require weight to be a maximum. Minimum weight generally means less material which leads to reduced production costs and economic advantages.

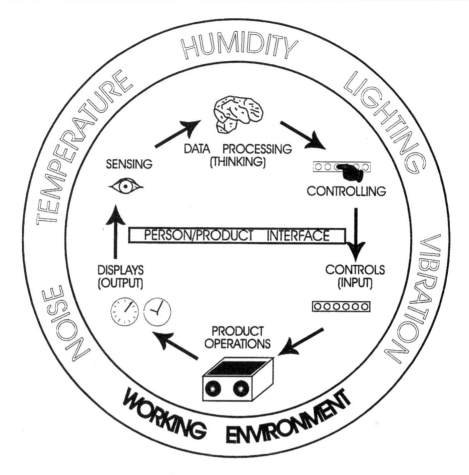

Figure 2.6 Person/product task division

Noise   The upper limits of noise levels which can be emitted by the type of product being designed should be specified. Regulations differ from one country to the next so either the standard which applies in a particular country or the lowest maximum limit amongst those countries targeted for export must be specified. These standards represent the maximum level of noise which is acceptable but lower levels could be specified, for example, to gain competitive advantage.

Manufacture requirements

Processes   The in-house manufacturing and forming facilities and the criteria under which external resources are sought should be specified. The required reliability of any source of supply and the required quality should be specified. Any special finishing processes which may be required should also be specified.

Materials   Materials for both the product and its packaging must be considered and the criteria governing the selection of materials specified without constraining the design team unnecessarily. The many criteria which must be considered are corrosion and wear resistance, flammability, density, hardness, texture, colour, aesthetics and recyclability. There are also many regulations governing the use of hazardous materials which must be included in the specification if relevant.

Assembly   The method of assembly should be specified; automatic, manual or assembly line. The rate of feed of components for assembly and time allowed are also important parameters. The specification should also contain statements with regard to the ease of disassembly.

Packing and shipment   The maximum size and weight for convenient transportation must be specified. Shape can also be important since stacking products together can reduce transport costs substantially. Provision of suitable packing, lifting points and locking or clamping of delicate assemblies should be specified to prevent damage during transport. It may also be important to ensure large products can be disassembled and reassembled easily for transport. The cost of packing and shipment must be added to the ex-works cost to ensure that the product remains competitive wherever it is used.

Quantity   The projected quantity of a product which will be sold can have a profound effect on the manufacturing methods and materials used. This must be specified as carefully as possible at the outset. This particularly influences the appropriate levels of tooling, with large quantities justifying expensive tooling.

Delivery date   It is important that realistic timescales are set for each stage of the design and production process. This is particularly important when a delivery date has been agreed with a customer and costly penalties for late delivery are built into the contract. Hence, the date by which each stage of the process is to be completed must be specified at the outset. The PDS of a single complex system which is to be designed and produced to an agreed contract will state dates by which the design, manufacture, erection, testing, commissioning and hand over of the fully working installation are to be completed.

Acceptance standards

Inspection   The degree of conformance to standards must be specified in accordance with relevant legislation and the objectives set in the PDS. The degree of conformance required to tolerances as stated within the rest of the specification must also be specified.

Testing   The methods of verification for the product should be specified along with the timescales for carrying out the necessary tests. It is usual on completion that acceptance tests are carried out in the presence of the customer. Tests often include safety interlocks, load capabilities such as speed and power consumption and reliability. Specified means and forms of testing should comply with standards where they exist. The PDS should contain a policy statement on the level of testing, such as every product to be tested or an agreed level of sample testing.

Standards   These may include national, international and company standards. There may also be many other rules, regulations and codes of practice which must be followed.

Patents Following a patent search it is important to state, and subsequently to ensure, that the design must not infringe any patents identified as being relevant. Patents are useful sources of information, particularly when you are beginning a new project with no previous experience in the particular field.

Disposal

Standards Individual country or international standards for disposal of products and materials must be listed in the PDS. The main implications should be stated. For example, most plastic materials used now must be identified during moulding of the component so that recycling and more importantly, reuse is made possible.

Legislation Any legislation governing the disposal of a product must be specified. Many governments are tightening their legislation with a view to ensuring recycling takes precedence over other methods of disposal, that manufacturers are responsible for accepting products from their last owners and that ease of dismantling and disposal are specified from the start. Also, legislation dictates that all materials used can be easily identified for subsequent recycling or disposal at the end of the life of the product. This must be specified.

Company policy Products which make less impact on the environment than similar products will have an increasing marketing advantage. They also afford a company significant advertising opportunities, which will also improve their competitive position. There are many ways of specifying this and only one is to specify increased life.

Hazards Any potential hazards that may cause difficulties at the end of a product's life should be identified and specified.

Operation requirements

Installation Where installation of a product is complex it should be specified. This is particularly important when small numbers of large devices are designed. The constraints should include construction, assembly, the time taken, provision of instructions and the skill levels required for installation.

Use The cost of ownership of a product, which should be minimized, is, in some cases, more important than the cost of initial purchase. Factors which influence this, such as the number of operators required, the skill level required from these operators, the cost of spares and the maximum tolerable energy consumption should be specified. Continuous, 24 hour a day, operation or the number of stop/starts in a relevant timescale should be specified. An alternative to dividing costs into separate categories is to specify a whole-life cost.

The power sources available should also be specified. These may include manual, gravitational, environmental, electrical, gas, water and internal combustion engines. Each should be specified exactly. For example, electrical power may be three-phase and 380–420 volts.

Maintenance A policy to minimize down time, simplify maintenance, ensure correct reassembly, provide easy access and provide interchangeable parts must be developed at the outset and specified. If there is to be any routine maintenance, service or overhauls

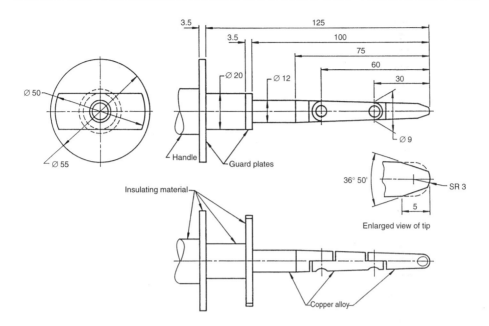

Figure 2.7 Test finger IV. From British Standard 3042:1971 (Extracts from BS 3042:1971 are reproduced with the permission of BSI under licence no. PD\1998 1956. Complete editions of the standards can be obtained by post from BSI Customer Services, 389 Chiswick High Road, London, W4 4AL)

the intervals and complexity of these should be specified. In order to simplify the maintenance procedure provision of special purpose tools and disassembly features should be specified if appropriate. The required skill levels of maintenance staff should also be specified. Guards should be easily removed. Levels of lubrication should be specified. An operation and maintenance manual must be supplied. Automatic lubrication should be considered.

Safety   There are many standards, a great deal of legislation and codes of practice which refer to all safety aspects of products. These should be listed in the PDS. As an example consider Fig. 2.7, which is extracted from British Standard 3042 and shows test finger IV. This is one of a series of probing devices for checking protection against mechanical, electrical and thermal hazards. Where standards do not exist it is normal to specify fail safe design with no sharp edges and that electrical panel isolators must be interlocked with the door, for example. Where headroom over walkways is less than 2 m suitable warning notices and head shock absorbers should be provided. Guards should be specified to eliminate danger to individuals or equipment.

2.3 Content of a PDS

As described, much work is required before an agreed or final draft of a PDS is produced. The content of each PDS will differ from any other but the way in which the information is ordered should always be the same. The essential information gathering which must precede the definition of the PDS is detailed in Chapter 9. Assuming the necessary information is available, including the identification of customers and any similar previous specifications, the complete format of the specification should be as follows:

(a) *Identification:* Title, designation, authority, date
(b) *Issue number:* Publication history, previous related specifications

(a)

(b)

Figure 2.8 (a) and (b) Excavator loader (Reproduced with kind permission of JCB)

(c) *Contents list:*   Guide to layout
(d) *Foreword:*   Reason for and circumstances under which the PDS is prepared
(e) *Introduction:*   Statement of objectives
(f) *Scope:*   Inclusions, exclusions, ranges and limits
(g) *Definitions:*   Special terms used
(h) *Body of PDS:*   Performance requirements, manufacture requirements, acceptance standards, disposal and operation requirements
(i) *Appendices:*   Examples
(j) *Index:*   Cross references
(k) *References:*   To national, international, or internal specifications.

Not all of the sections which represent the full format are necessary in every case. For example, a foreword should only be included where it would assist the understanding of the PDS. Also, students will find the identification of issue numbers and authority fatuous on many occasions since they are only relevant when working within a company environment. However, just as it is important with detail drawings and components to identify unique identifying numbers so it is with specifications. You should therefore identify a PDS as fully as is possible in the circumstances.

2.4 Sample PDS

There are many specifications which run to multiple volumes and include such things as contractual and warranty agreements. It is not possible, nor desirable, to produce such a PDS here, so the example presented contains the level of detail considered appropriate for you as students to produce during your engineering course. The PDS is for a suspension mechanism to isolate the vibrations of an excavator loader, such as the one in the photographs in Fig. 2.8 (a) and (b), from the operator.

ISSUE: 3	PRODUCT DESIGN SPECIFICATION for	REFERENCE NO.: PDS014
DATE: 17:06:98	UNDER SEAT SUSPENSION UNIT	
RELATED SPECIFICATIONS:		
ISSUING AUTHORITY:		
CONTENTS:		

FOREWORD: The photograph in Fig. 2.8 shows an excavator loader. Historically, prior to 1975, machines of this type did not have any suspension, other than that provided by the pneumatic tyres. Investigations into injuries sustained by operators of farm tractors and off highway construction vehicles, like the excavator loader, have revealed a high incidence of back trouble. These injuries are more marked in operators of farm tractors because of the need to observe attachments being towed at the rear of the machine. This means that the spinal column is twisted and the induced stresses are consequently higher.

INTRODUCTION: The objectives are:
To design a suspension mechanism for the operator's seat.
To design the mechanism in such a way that the position and orientation of the seat is fully adjustable.
To design the mechanism so that driver vibrations are damped to within acceptable limits.

SCOPE: All machines, whether with a fully enclosed cab or a simple canopy, are to be provided with such a suspension mechanism.

DEFINITIONS:

PERFORMANCE REQUIREMENTS:

The mechanism must allow full adjustment of the seat position. To comply with ISO 4253 these adjustments are rotate through 180 degrees in either direction, 80 mm up and down in the vertical plane and 150 mm front and back in the horizontal plane.

Increments of adjustment must be less than 30 degrees and 25 mm respectively.

The natural frequency of vibration of the combined seat and operator must be <2.5 Hz. Isolation criteria for class 3 seats as set by ISO 7096.

The mechanism must still operate with the machine on a 30 degree slope in any direction.

Suspension travel must be vertical and a maximum of 110 mm. Amplitudes must be limited under resonant conditions and step inputs.

The temperature range during operation is between −10 and +50°C and whilst stored could drop to −30°C.

The humidity will range from 0 to 80%.

The suspension mechanism will also be subjected to rain, snow and heavy organic and mineral grime.

The ex-works cost of the mechanism must be <£30.

The target population of operators is to be restricted to people between the ages of 19 and 65. Sizes, weights and strengths are to be between the 5th and 95th percentiles. For example, adjustment must accommodate drivers in the weight range of 60 to 130 kg.

The quality of the mechanism must be consistent with the rest of the machine.

The required design life is 10 000 hours of operation.

The required reliability is 90% over the 10 000 hours of operation.

The appearance must be as rugged as the rest of the machine.

The weight of the complete mechanism must be <50 kg.

The maximum overall size is $0.5 \times 0.5 \times 0.5$m. In the horizontal plane the mechanism must have a radius about the centre of rotation <300 mm.

The mechanism must be capable of being fitted to the full range of seat bases and machine floors.

MANUFACTURE REQUIREMENTS:

The machine will be assembled on a ten stage assembly line. The mechanism will be assembled prior to installation as far as is possible. Installation must take <20 minutes.

The mechanism is to be manufactured and finished in-house.

Any materials can be used as long as they comply with other statements in this specification. 6000 are to be produced each year.

ACCEPTANCE STANDARDS:

In accordance with ISO 3776 anchorage points must be provided for seat belts which accept a pull load through the suspension of 5000 lbs.

Every mechanism will be inspected prior to assembly in the machine.

Accelerated cyclic tests of five fully loaded mechanisms are to be carried out to verify the reliability levels and fatigue strength.

The mechanism must not conflict with existing patents.

DISPOSAL:

The suspension mechanism must not contain any hazardous materials and all polymeric materials used must be clearly identified.

OPERATION REQUIREMENTS:

Adjustment of the seat position or the level of damping must be easily carried out by the operator whilst in the sitting position in <30 seconds.

Removal of the mechanism from the machine by one person must be possible in <30 minutes.

The device is to be maintenance free for the life of the machine.

Secure locking must be provided after adjustment.

Seat movement and locking must be fail safe.

It must not be possible for the operator to trap their fingers in the mechanism.

2.5 Principles

Specification principles

Definition   With the agreement of the customer all important technical aspects of the future 'product' must be specified.

Information   The specification must be informed by relevant and up-to-date information gathered from a wide variety of sources.

Function   A clear statement of the function(s) the product is required to fulfil is the starting point of the specification.

Constraint   The many aspects of the product which the customer and market surveys indicate are required must be quantified into statements that the engineering team can work towards.

Iteration   At the outset the specification can only be regarded as a draft. As the project progresses more information will surface which will add to or contradict the original draft. This is normal and acceptable, as long as the 'customer' agrees to the changes.

2.6 Exercises

1. Write a detailed specification for a car which is to be used in towns.

2. Write a specification for a foot-pump to be designed for pumping up car tyres.

3. A market opportunity exists for a can opener which is operated by disabled people with only one hand. The device is to open the can in such a manner that the contents are not spilt or contaminated, the contents can be easily removed and no dangerous or jagged edges are left exposed. Write a specification for the device.

4. With the increasing popularity of the game of snooker, most UK towns and cities now have many snooker and pool halls. There is a substantial market for a device which will make the cleaning of the tables easier and quicker. The present cleaning procedure for a snooker table is as follows:

   (a) Brush debris out from under the cushion into the middle of the table.
   (b) Use hand brush and dustpan to remove debris, by brushing up the table to baulk.
   (c) Iron up the table to baulk.

   Write a specification for a combined vacuum cleaner and iron which would reduce the time taken in cleaning a snooker or pool table.

5. Shown in the sketch at the top of p. 30 is an idea for a working platform. The principle is to span two ladders with support bars pushed through the hollow rungs and to use these bars to support the platform. Write a specification for such a platform.

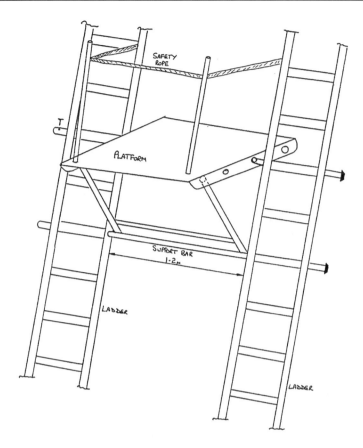

You can assume that the idea is sufficiently innovative to be patented. Would you recommend that a company produce such a device? Write a short report explaining your reasons for recommending continuation or rejection of the project.

6. Current multi-gyms found in health spas, leisure centres and gyms are extremely large, taking up a lot of space. A market survey has been conducted indicating a sizeable market for a similar device, if it were to be available for home use. The new design must incorporate all the basic movements offered by large multi-gyms, but be small enough to fit into a spare room or garage of a typical house. The movements offered should include shoulder press, chest press, leg raisers, leg/thigh extension, leg curl, latissimus pulldown, triceps pulldown and biceps curl.

    Produce a detail specification for a small multi-gym. You should assume that the specification will be passed to another designer who will not be able to ask for further information.

7. Due to increasing demand for high-speed public transport, particularly on the railways, the requirements for more precise rail-track performance became a necessity. This led to the introduction of longer rails and continuously welded tracks. Problems associated with track expansion were overcome by using sleepers at shorter spacings and clamping the rail to them. This causes compression stresses in the rails, but does not allow expansion.

During routine inspection of the rails it was discovered that uneven wear occurred and this was especially the case where the track was curved. The worn lengths had to be replaced, not necessarily with the full length of track. The procedure adopted involved flame-cutting the section out, replacing the section and then welding back together. Railtrack have identified a problem with these welds because they fatigue and crack after a short period of time. This is due mainly to the poor control of the cutting operation and not due to poor weld quality. Therefore, a portable 'Rail-Cutter' is required which could be used on site and track locations and be capable of cutting any rail section to such a high standard that no further attention of the rail ends is required.

Produce a detailed specification for such a 'Rail-Cutter', broken down into functions and constraints.

8. An ex-seagoing officer has had an idea which he claims would substantially reduce the cleaning time for the holds of large bulk carrying ships. The idea is a portable bilge strainer which would prevent solids entering the bilges and damaging the bilge pumps during cleaning. These solids can be anything from coal to grain and the current washing procedure is outlined below:

(a) Inspection of bilge, strums and roses prior to washing.
(b) Rough sweep of all accessible areas and removal of solids from hold prior to washing since the solids are much lighter when dry.
(c) Fit perforated grids over bilge well. Previous cargo and state of the hold will dictate the grade of fine netting, wire gauze or perforated sheet used.
(d) Use of pressure hoses capable of reaching all areas of the hold. Someone in charge with communications link to the bridge and engine room.
(e) Monitor amount of water in the hold and washing should cease if bilge stops pumping or cannot prevent water build up.
(f) Remove solids from base of hold and bilges. Clean and inspect the bilge.

Unfortunately, the theoretical procedure outlined is seldom, if ever, this straight forward. Invariably the pumping out procedure slows due to a combination of problems such as choked bilge cover plate or pump. One remedy is to remove the bilge cover plate but this allows solids to enter the bilge. The subsequent removal of solids from the bilge is a slow, awkward and unpleasant task as there is hardly room for one man to enter the bilge.

The aim is to prevent the necessity for the removal of the cover plate with all its associated problems. Basically the idea involves using a top hat shaped filter standing on the hold floor and covering the bilge well. The sizes of the bilge well openings vary from 450 mm diameter to 900 mm square with 100 mm radiused corners. Space available for the filter is restricted to a radius of 700 mm from the centre of the well. The thickness of the hold base is between 20 mm and 30 mm and can undulate by 100 mm. The depth of the bilge is 1.2 m. A typical hold is 20 m deep and contains 10 bilge wells. The maximum force imparted by a hose is 300 N and hosing down must take place in the open sea.

Write a specification for the bilge strainer based on the design brief outlined.

Creativity

Included in this chapter are many of the methods which can be used by individuals or groups in order to increase their creativity and obtain potential solutions to problems. These methods are illustrated by examples and include lateral thinking, avoiding 'set', inversion, analogy, empathy, fantasy, free-wheeling, brainstorming, morphology and synthesis. The chapter finishes with presentation techniques for concepts, illustrated by the on-going example of the seat suspension mechanism.

3.1 Introduction

Having made a thorough identification of the problem and written a PDS defining the boundaries of that problem, however incomplete, the next stage in the design process involves exploring these boundaries. This is the divergent stage of the design process and involves the generation of as many concepts as possible with the potential for solving the problem. This is the most creative stage of the design process and the techniques described can be applied equally well to both completely new product concepts and to those which only involve development of existing designs.

Society develops rules of behaviour which are considered normal and with which most adults readily conform. In general, education systems, right from the earliest stages, encourage conformity and discourage creativity and invention. Pre-school children have vivid imaginations which are often suppressed by rules such as those of mathematics and language. We were all potential entrepreneurs at the age of four! It is commonly academic success in the application of vertical thinking, particularly in mathematics and science based subjects which leads eventually to engineering as a career. Perhaps more worryingly we are taught in the numerate subjects, such as mathematics and the sciences that most problems have one unique answer. In engineering this is seldom, if ever, the case and in design we are continually searching for an optimum or compromise solution.

Vertical thinking is best explained by means of an example. Consider the story of how monkeys are caught by burying a narrow mouthed jar of nuts in the ground. A monkey comes along, sees the jar, puts its paw into the jar and grabs a handful of nuts. The mouth of the jar is of such a size that it admits the unclenched and empty fist but is not sufficiently large for a clenched fist full of nuts to be removed. The monkey is unwilling to release the nuts and is therefore trapped. With vertical thinking the obvious way of looking at a situation is grasped, perhaps because it has proved useful in the past. Once it is grasped there is a reluctance to let go. The suggestion is not that vertical thinking must be avoided but that it must be complemented by an attempt to escape from a particular way of looking at a situation.

Such vertical thought processes are essential in most engineering specialisms and discipline is essential in detail design work where limits and fits, drawing standards and analysis rules must be followed. However, the basic thought process employed in generating ideas during the concept stage should be that of lateral thinking. In vertical thinking information is used for its own sake in order to progress to a solution, whereas in lateral thinking, information is used, not for its own sake, but provocatively to bring about repatterning. The main purpose in employing lateral thinking is to challenge all assumptions and to try and restructure any pattern. General agreement regarding the continued validity of any assumption is no guarantee that it is correct. It is historical continuity that maintains most assumptions, not a repeated assessment of their validity. Consider the following problem.

```
    •     •     •


    •     •     •


    •     •     •
```

Nine dots are arranged as shown. The problem is to link up these nine dots using only four straight lines which must follow on without raising the pencil from the paper. (The answer is at the end of the chapter.)

The implication of the foregoing is that personal creativity can be improved by forced application of techniques such as lateral thinking. However, some limitations must be admitted since heredity, environment and past education impose restrictions on inventive ability. Also, for design engineers, creative ability depends to a large extent on a thorough knowledge of scientific and technical principles. Having accepted these limits it is undoubtedly true that creative ability can be improved by the application of the techniques described. These techniques are proven, but their successful use requires effort and practice.

Anything which improves creativity, such as doing cryptic crossword puzzles, playing chess or attempting creative growth games is to be encouraged. As an example, how many answers are there to the question what is half of thirteen? I propose six and you may be able to think of more. In creative problem solving, it is more important to look at the problem from different vantage points rather than run with the first solution which comes to mind. The six answers to half of thirteen can be found at the end of the chapter, with explanations.

Psychologists have many scales with which they categorize people into types. One such scale is the judging–perceptive scale. At one extreme is the judging person who when confronted with a new situation quickly judges it good, bad or how it ought to be. Specialization tends to bring this about. The expert is a person who within a certain field is best qualified to distinguish right from wrong and good from poor. At the other end of the scale lies the perceptive person who is more concerned with how things are and how they work. To become more creative a person should practice being less judging and more perceptive. A course in art is very helpful in increasing visual perception. Improving perceptiveness requires continual reminding at first since it really involves a change in personality, but it can be done.

One sure way of improving engineering creativity is to carry a note pad everywhere and to sketch any interesting features of existing products which you come across. Where observation is concerned chance favours the prepared mind!

Creativity can be improved, but only with hard work and concentration. The motivation must be to succeed and the working environment must be so arranged that creative thought is encouraged. According to Eddison, invention is 95% perspiration and 5% inspiration. Eddison should know since he is claimed to have tested over 6000 materials before discovering a particular species of bamboo suitable for the filament of an incandescent lamp.

In an effort to define the creative process several inventive people were asked to review their own behaviour. A summary of these descriptions gives the creative process as:

- *Preparation:* information gathering, formulation of the problem
- *Concentrated effort:* application of creativity techniques
- *Withdrawal:* period of mental rest/incubation away from the problem
- *Insight:* the concept which is the solution
- *Follow through:* generalizing and evaluating

Time for creativity is generally not planned in manufacturing companies because schedules have to be met and products launched into the market place as quickly as is possible. However, if inventive solutions are expected then work must be planned so that the inventive process may flourish. More particularly, the period of withdrawal identified as most important by the people canvassed must be positively encouraged.

There is another reason why more time and resources should be allocated to the early design stages and this is illustrated in Fig. 3.1(a) and (b). The graph in Fig. 3.1(a) shows a

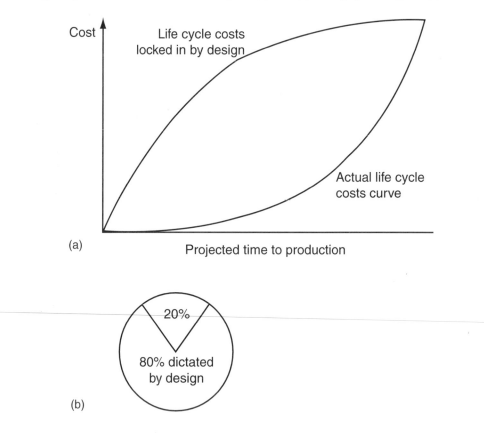

(a)

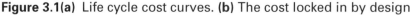

(b)

Figure 3.1(a) Life cycle cost curves. **(b)** The cost locked in by design

typical product cost to production curve over, for example, four years. The manufacturing costs locked in by design rise steeply in the very early stages. However, the actual costs incurred by the company, illustrated by the lower curve, show that very few resources are applied until much later in the project.

Extra time allowed during the design stages of a project is invariably rewarded since, as illustrated in the pie chart of Fig. 3.1(b), approximately 80% of the cost of manufacturing a product is locked in during these early stages. Any remedial action taken after the design is finalized and production runs begun can only have a minimal effect on manufacturing costs and profit margins.

It is almost always the case that during the creative stage questions will arise which will necessitate alteration or extension of the PDS. These points should be noted immediately and a new issue of the PDS created.

3.2 Psychological 'set'

Probably the single most important obstacle to inventiveness is what the psychologists call 'set'. This means a predisposition to a particular mode of thought. To illustrate 'set' the reader is asked to solve the problem shown in Fig. 3.2. Let us assume that engine oil is

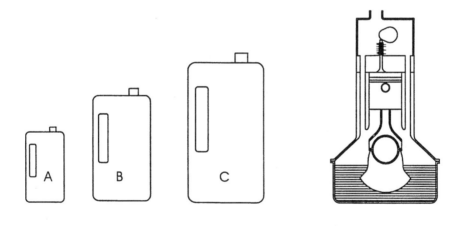

Problem	A	B	C	Desired	Answer
1	5	7	10	8	C − B + A
2	21	23	26	24	
3	6	19	25	12	
4	7	10	18	15	
5	6	8	11	9	
6	2	6	10	6	
7	8	10	13	11	
8	9	23	18	4	
9	17	34	34	14	
10	14	28	23	9	

Figure 3.2 Problems to illustrate set thinking. For each problem, what is the simplest way to obtain the desired quantity, using full oil cans?

supplied in three different sized containers, A, B, and C, that no partial volumes are marked on the containers and that the same sizes of empty containers are available. Exact quantities are to be used to fill the engine with oil to a desired, though unmarked, level. There are ten problems and as an example problem 1 has been solved. The desired volume of 8 units can be achieved by pouring A into B until B is full, pouring the residue of A into the engine and adding C, i.e. A – B + C. The reader should now attempt to solve the remaining problems as quickly as possible. The answers are to be found at the end of the chapter.

Being 'set' on a particular method or solution is either developed by habit or part of personality. Perseverance, which is generally to be admired, can easily become stubbornness! In such cases the aim can become to make a particular solution work rather than investigating alternatives. In critical situations, old methods are clung to more and more. Familiarity encourages 'set', which is one of the main reasons for the increasing use by companies of outside consultants. Outside consultants are not bound by previous knowledge or history, can challenge traditional approaches and provide a fresh mind in the solution of the problem.

'Set' also affects learning. Some things have to be believed to be seen! Many experiments clearly show that people learn more facts which support their opinions than they do facts which contradict these opinions.

Consider the example of a hot water bottle used for warming a bed. For many years the design has remained 'static', except for the use of better materials as they were developed. The main problem, the potential for leakage causing either a scalded foot or wet bed, has been largely solved by more and more ingenious stoppers. However, it may be that the very name of the product has inhibited radical development. Engineering designers have to be very aware of traditional approaches and question what the customer requires. In this case it is simply a device to warm a bed. Water need not be used nor is a bottle necessary. A questioning approach recently led to the development of a sealed gel-filled bag which is warmed in a microwave oven. This neat solution overcomes all leakage problems but could only stem from overcoming 'set'.

Such a questioning approach led to the design of the JCB Fastrac shown in the photograph of Fig. 3.3 along with a traditional farm tractor, Fig. 3.4. Market analysis showed that a farm tractor spends most of its working life towing trailers along roads and not working in the fields. Traditional tractors are designed for working in the fields not road work. The

Figure 3.3 JCB Fastrac (Reproduced with kind permission of JCB)

Figure 3.4 Traditional farm tractor

Fastrac is designed to travel at road speeds, so as not to hold up the traffic, as well as to be 30% faster when tilling the land. This is achieved by even weight distribution over four equally large wheels and a unique suspension system. History alone will reveal the level of success of this radical new design.

3.3 Inversion

This is a deliberate method for breaking out of 'set' thinking which involves viewing a problem from a different angle or stand point. If we are looking at a problem externally then forced consideration from the inside is inversion. The use of the following words, and many more, has been shown to stimulate ideas:

Adapt	Expand	Magnify	Re-arrange
Modify	Reduce	Reverse	Substitute

An example of reversal is Aesop's fable of the water in the jug which was at too low a level for the bird to drink. The bird tried in vain to obtain a drink but could not succeed as long as he only considered taking water out of the jug. Once the situation was reversed and he thought of adding something to the jug the solution was obvious. He dropped pebbles into the jug until the water level rose sufficiently for him to drink.

A similar story concerns an experimental machine for coal mining applications. The machine operated for two months without any problems. Then it was noticed that the bucket on the machine, which was lifted by hydraulic ram, would not lift the required load. A team of development engineers was initially baffled as to the cause of the problem. Then they looked in the hydraulic tank, which the sight glass indicated was full, and found it almost full to the top with coal. The instruction on the tank said maintain the level of hydraulic oil at a particular level, and because it was almost a five mile trip back to the oil store at the pit head, the machine operator had added rocks, much like the bird, to maintain the level. Obviously this caused the loss of power and emphasizes the need for operator proof design and clear and unambiguous instructions.

An example of inversion applied to engineering design concerns a cab for a new earth-moving machine. Every new design of cab has to meet strict Falling Objects Protection (FOPS-ISO 3449) and Roll Over Protection (ROPS-ISO 3471) standards. These standards set out tests for a cab, during which no part of the frame must encroach on the driver envelope. As shown in Fig. 3.5(a) the particular cab in question failed one part of the test, that where a large weight is swung at the top edge of the cab. The cab failed mainly because the fixings holding the cab to the machine base sheared off and the cab moved bodily sideways.

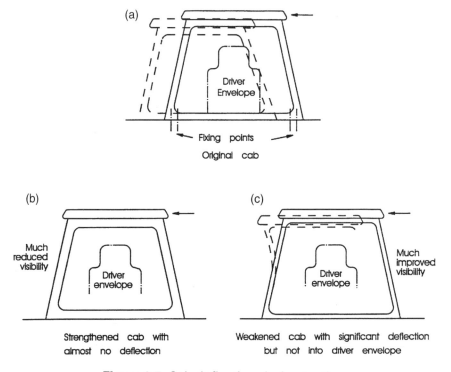

Figure 3.5 Cab deflection during testing

The instinctive reaction was to re-design the cab with strengthened sections and fixings. This was very disappointing since it was inevitable that visibility would be restricted by more metalwork and less glass. However, careful study of the mode of failure and the use of inversion suggested there was an alternative course of action. Instead of strengthening the cab it was suggested that it was already too strong and that the problem during the test was that the loads were transmitted through the frame directly to the fixings. If the frame was weakened this would absorb some of the load by deflection and lessen the load on the fixings.

The scheduled launch of the machine could not be put back without severe financial losses so it was critical that a new cab was designed and tested quickly. Two new cabs were designed. The first with large gussets in the corners, larger fixings and stronger sections as illustrated in Fig. 3.5(b), and a second cab with weaker sections as in Fig. 3.5(c).

Both passed the test and the weaker design with considerably better visibility and lower manufacturing costs went into production.

3.4 Analogy

Another way of generating concepts is to break down a problem into small parts and then consider analogous problems and their solutions from either within engineering or outside. One example of this would be the application of basic kinematic principles to the solution of mechanical design problems. This could be accomplished by studying a list of standard mechanisms to ensure that no good solutions are overlooked. Once a mechanism or linkage is selected to serve as the foundation for the device, features can be added or adapted to meet the constraints of the particular problem at hand.

To use analogy we must develop a thorough knowledge of the way things work generally and have an understanding of other disciplines such as biology, physiology and psychology. One often very fruitful analogy when designing machinery is to imagine how a person would perform the task and attempt to emulate this in the design. This could also be termed employing empathy for problem solving, since empathy involves becoming the part and seeing things as the item or artefact we are considering.

A further useful source of ideas comes from analogy with nature. Nature is a powerful solver of a wide range of problems and many modern engineering inventions replicate nature. For example, consider the following:

- As a bat flies (Fig. 3.6) it emits sharp cries which bounce off obstructions and warn the bat of their position. In other words the bat uses *sonar*. The bat sonar is an amazing discriminator: in a bat-swarm, in cave or night air, a bat can know its own sound among thousands of mobile neighbours, detecting its own signals even if they are 2000 times fainter than background noises. It can 'see' prey, such as a fruit-fly, up to 30 metres away by echo-location, and can catch four or five in a second. This whole auditory system weighs a fraction of a gram! Gram for gram, watt for watt, it is millions of times more efficient and more sensitive than the radars and sonars contrived by man.
- To jet-propel itself for a high-speed swim, the squid (Fig. 3.7) sucks water into its body, then shoots the water out of a tube under its head. The squid uses jet propulsion.
- The idea of producing a flying machine that spins was originally thought of many thousands of years ago in China when a toy based on a flying top was produced. This idea probably came from watching sycamore seeds fall from the trees.
- The caribou has wide feet, or snow shoes.

Figure 3.6 Bat in flight

Figure 3.7 Squid swims by jet propulsion

Figure 3.8 Scorpion with sting in tail

- The scorpion (Fig. 3.8) applies its sting through a device which we now know as a hypodermic needle. Here the scorpion is injecting a locust.
- When biting, a snake applies an anaesthetic which lessens the pain of the victim. This principle has only relatively recently been used for operations.
- Sea snails cling to rocks by means of suction. Suction cups are in wide use today.
- Birds brake with their tail feathers just as planes do with flaps.

3.5 Fantasy

Insight into the truly creative thought processes can be gained from a study of modern and classical literature. Those with fertile minds, such as Wells, Huxley, Clarke and Asimov, are well known for their projected visions of the future. Many inventors have used ideas contained in these science fiction books to bring about a solution to an intractable problem.

As specific examples consider *20,000 Leagues Under the Sea* by Jules Verne (1870) and *Frankenstein* by Mary Shelley (1818). In the first the then almost unheard of submarine is extended into the sophisticated device roaming the seas of the present day. Perhaps more inventively that submarine had electric lighting and the book contained quite a detailed section covering the chemical oxygen production necessary to keep the craft submerged for long periods. In the second, the notion of electric charges giving life is introduced, in much the same way as electric shock treatment of cardiac arrest victims is used by medics today.

What is being suggested then is that a study of the more modern writers of science fiction could provide an increased understanding of the creative process and may even reveal potential inventions in the making. One worrying prediction concerns *I, Robot* by Asimov (1950) where robot brains program themselves into the position of absolute master!

3.6 Technological advances

Many 'scientific' advances are made each year which affect the work of engineering designers and provide exciting opportunities for product improvement or the introduction of new products. As an example, new materials are being developed almost daily. Thus, it is essential that professional engineers engage in some form of Continuing Professional Development (CPD), such as reading trade magazines and attending updating courses.

Consider the humble vacuum cleaner. This was, to all intents and purposes a fully developed or 'static' product. Then along came the application of new technology and

Figure 3.9 Dyson Dual Cyclone vacuum cleaner (Reproduced with kind permission of Dyson Appliances Ltd.)

materials, in the shape of the Dyson, illustrated in the photograph of Fig. 3.9. The Dyson vacuum cleaner has Dual Cyclone filtration and many other patented design features. The main advantages claimed over conventional vacuum cleaners are that no bag means 100% suction, 100% of the time and no bag odour. The vacuum cleaner has the largest electrostatic filters found on any vacuum cleaner, is designed to sit on the stair, has integral tools, the large rear wheel/small front castor design of the Dyson makes it easy to manoeuvre around corners. The body of the Dyson Dual Cyclone is made from ABS and polycarbonate.

3.7 Brainstorming

The problems faced by engineering designers are many and varied, ranging from those necessitating radically new solutions to the everyday problems such as oil seal, bearing and gear design or selection. The full range of problem types can benefit from the application of techniques which are often categorized as 'organized ideation', such as Brainstorming.

Brainstorming involves attacking a problem with the full creative power of the brain and is normally a group activity. The basic principle is that of association of ideas. Everyone has experienced the situation where an idea voiced by someone else brought a thought to mind or where a word spoken during a conversation makes you think of something that would never have come to mind otherwise. In other words ideas are stimulated by the ideas put forward by others.

Brainstorming as a technique was first suggested in *Applied imagination* by Alex F. Osborn (1953) as an alternative to the more usual business meeting. He regarded such meetings as a waste of time since they often did not yield anything of value. Osborn set out four rules to be observed by a group discussing a problem:

Criticism is ruled out   Evaluation and criticism of ideas at this stage must be avoided since it inhibits the production of ideas. Even the wildest idea can have some usefulness. The major inhibitor to successful brainstorming is one of attitude and all judgement must be deferred.

Freewheeling is wanted   Give free rein to thoughts and creative imagination.

Quantity is wanted   The basic premise here is that quantity breeds quality. The more concepts that can be produced the more likely it is that a good solution will be found.

Combination and improvement are sought   After the initial flood of ideas each is examined to ensure that the underlying principle is clearly identified. More ideas may emerge as a result. Where possible ideas are combined.

During a brainstorming session all suggestions and ideas are recorded. This is done until many ideas have been collected and current thinking suggests that not until 70 concepts have been generated should the process be halted. Recording ideas as they arise means that the memory is not being relied upon, the mind remaining uncluttered and free to produce additional ideas.

The optimum duration of a brainstorming session is half an hour, with most useful ideas being suggested in the second quarter. Normally in a session, individuals are free to contribute at any time but when a group is over six in size this is not practical. Stein, in his excellent text, *Stimulating Creativity*, favours sequential brainstorming for all relatively large groups. During this the group members sit in a circle and every member in turn puts forward one idea or suggestion.

One further variation of brainstorming technique is worthy of mention since it has proved most fruitful when used by students of engineering design. This is a forced creativity exercise in which the group as a whole is subdivided into groups of three. Everyone considers the same problem and must write down a suggested solution in one sentence or simple diagram in one minute. The papers are folded to hide the first solution and then passed around the smaller groups twice and new solutions written down. Each piece of paper contains six ideas and there are the same number of papers as participants. Therefore many different ideas can be generated in little over six minutes. There are of course many permutations on this theme.

Many companies are now introducing total quality management procedures and as part of this process hold regular brainstorming sessions. The personnel involved in these meetings are not limited to design staff, nor indeed to technical staff. All staff are considered valuable in the process of generating concepts and design engineers must accept this fact for the overall good of the company.

3.8 Morphological analysis

When a device or system being designed must satisfy several functions or combine several features it is worth subdividing the problem. Concepts are then generated to satisfy each smaller problem area and then later combined. In order that every potential combination of these concepts is considered then a morphological analysis should be carried out and a morphological chart drawn up. A four stage approach is recommended.

(1) Make a close examination of the specification and list the functions and features which are required.
(2) Identify as many ways as possible of providing each feature or function.
(3) Draw up a chart with the essential features or functions on the vertical axis. Along each row enter the means of achieving each of the functions or providing each desirable feature.
(4) Identify all practical combinations which satisfy all requirements of the whole.

Morphological analysis is best illustrated by means of a case study. Consider, as an example of morphological analysis, the problem of digging small trenches. These trenches could be required for various reasons; drainage channels in a large lawn or playing field or for carrying services on a new housing estate, for example. The problem is to design a new device, a mini-trencher, which would dig a relatively shallow trench more cheaply and quickly than two labourers using spades. Taking the four steps as outlined we must first list the essential features:

• support preventing sinking in soft ground;
• provide forward motion;
• type of power source;
• excavating mechanism used;
• power transmission system;

- stopping the mini-trencher;
- spoil removal to side of trench;
- position of the operator if required.

In the next step alternative concepts are generated for the provision of each feature. Each full or part concept may be presented as a simple sketch or in words, depending on the complexity of the concept. In this case the concepts generated for the sub-problem of excavating mechanism are back hoe, plough, auger, dredger type buckets and wheel with buckets. All other requirements were considered in a similar manner and a morphological chart created with the concepts generated along the horizontal axis, as can be seen in Fig. 3.10.

Feature	Solutions				
Support type	wheels	tracks	skis	*balloon tyres*	
Forward motion	drive wheels	*single wheel drive*	tracks	spiked wheel	manual
Power source	*petrol*	diesel	electric	bottled gas	manual
Excavating mechanism	back hoe	plough	auger	**dredger buckets**	wheel buckets
Power transmission	hydraulic	chains	belts	*gears and shafts*	
Stopping trencher	brakes	dead man's handle	clutch	*torque limiter*	
Spoil removal	chute	auger	*plough*	conveyor	
Operator position	seated at front	seated behind	*walking behind*	remote control	

Figure 3.10 Morphological chart for mini-trencher

All that remains is to identify all the possible combinations of features which may satisfy the overall requirements. As will be self-evident an almost infinite number of 'new ideas' could be generated by such an analysis. The sketch in Fig. 3.11 represents the combination identified on the morphological chart of Fig. 3.10 by the emboldened and italicized ideas. That is balloon type tyres, single drive wheel, petrol engine, ladder boom with dredger type buckets, gear box transmission, torque limiter, plough spoil removal and operator walking behind. It is likely that this would be one of the preferred options but the selection procedures contained in Chapter 4 should be used to confirm this.

3.9 Presentation

The three-dimensional sketch of the mini-trencher in Fig. 3.11 is far too detailed for the form of presentation necessary at the concept stage. Indeed, too much detail is counter

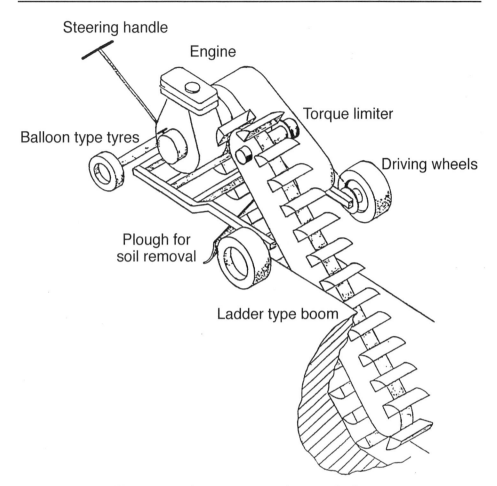

Figure 3.11 Concept sketch for trench digger

productive since it can detract from the working principles of the concepts. The selected concept is refined and detailed at a later stage, but most of the effort in developing each design beyond simple sketches is wasted since the majority of concepts will be discarded.

What is very important when presenting concepts is that each concept generated is given equal importance by using a standard form of presentation. The form recommended is that of concept sketches or line diagrams. In many engineering disciplines standard symbols have been developed, particularly for circuit diagrams. Hydraulic and pneumatic components, electronic components and mechanical components all have standard simplified representations.

Consider the seat suspension mechanism problem for which a specification was written in Chapter 2. Many concepts were generated and sketched and after employing the concept generation techniques contained in the current chapter along with the combinational methods described the overall concepts shown in Fig. 3.12 were produced. Six only are illustrated since by inspection the other concepts had serious and obvious deficiencies making them impractical. These concept sketches are sufficient for concept selection purposes and should be accompanied by explanatory notes as necessary.

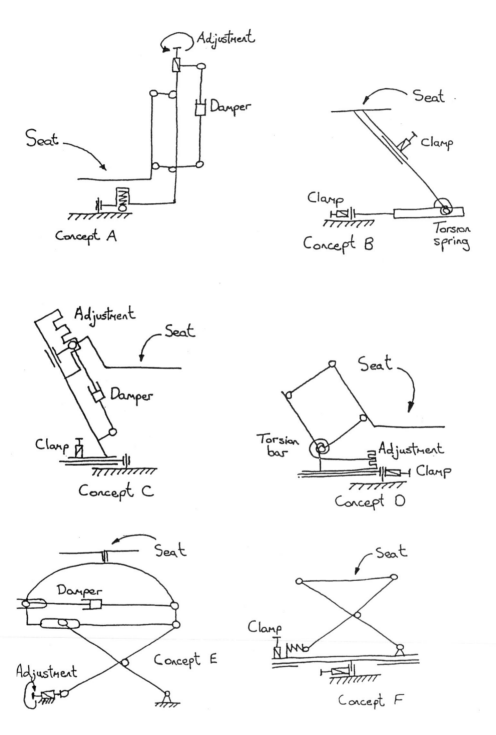

Figure 3.12 Concept sketches for seat suspension mechanism

3.10 Answers to problems

1. The nine dots can be joined as shown. The false assumption usually preventing you from obtaining a solution is that the straight lines must link up the dots without extending beyond the boundaries set by the outer lines of dots. This restriction, whilst often assumed is not stated in the 'design brief' or 'problem specification'. Assumptions, tradition and hearsay should always be questioned.

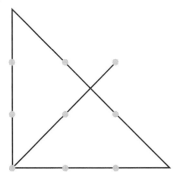

2. What is half of thirteen?

6.5 is 13 divided by 2. 13 consists of a 1 and a 3. Therefore 1 and 3 both make up half the number 13. In roman numerals 13 is written XIII. A horizontal line dividing the letters in half reveals VIII on top, which is 8. XIII divided vertically in half gives XI/II so both 11 and 2 are half of thirteen.

The answers are 1, 2, 3, 6.5, 8, 11.

3. If in the engine oil pouring problem you obtained the answers for all problems of A − B + C then you were a victim of 'set'. However, if you obtained the correct answers to 6, (B) and 10, (C − A) and you solved the nine problems in under 2 minutes then you were not 'set' by the other answers. It is interesting to note that if the problem was inverted and worked from the bottom up then correct answers would be obtained more readily.

3.11 Principles

Concepts generation principles

Divergence   The concept stage of the design process begins with the generation of many potential solutions by broadening of the problem.

Creation   Many potential concepts are created by teams using such methods as brainstorming.

Inversion   Different concepts can often be created by the simple technique of looking from a different angle. For example, imagining yourself to be the device to be designed.

Analogy   Nature in many forms has solved problems and the methods employed can often be modified in the solution of engineering problems. Personal analogy is a powerful technique.

Fantasy   In this technique the imagination runs free and concepts are accepted without criticism. 'Freewheeling' is sought.

Combination   Having created many concepts, the possibility of combining aspects of these concepts to create an optimum solution is investigated.

Observation   Creative people take regular note of their surroundings and are more perceptive than average.

Gestation   The creative process, following a period of concerted effort, requires a period of withdrawal from the problem.

Iteration   During the generation of concepts it is inevitable that the boundaries of the PDS will be questioned. Earlier stages of the design process may need to be revisited.

3.12 Exercises

1. Your company has just bought the salvage rights to the wreck of the *Titanic* as a speculative venture. The wreck is lying at a depth of approximately 2 miles, is standing on end and at a pressure of 300 bar and temperature of 4°C. Use group brainstorming techniques to generate concepts for raising the *Titanic*.

2. Blind people have difficulty in filling receptacles with hot liquids to the required level. Problem areas are the water level in a pan when cooking, liquid levels in a cup and the level of water in a bath. Generate as many concepts as possible for a simple hand-held device which will indicate to the person the level of liquid in all these cases.

3. Generate concepts for replacing the traditional type of scarecrow with alternative methods of bird scaring.

4. As crime figures mount, fear of strangers coming to the door, particularly at night, leads to many people refusing to open the door unless they are sure who is calling. A security chain is good, but could quickly be snipped through with bolt cutters. Remote TV surveillance and electronically operated locks are relatively expensive. Using group brainstorming techniques generate as many concepts as possible for the design of a simple low-cost means of preventing a front door from opening fully which can be easily and quickly disengaged if the caller turns out to be a friend.

5. Within a large catering organisation many slices of bread are buttered. This is a tedious and time consuming task. Generate six concepts for automating the process of applying the butter to the bread.

6. Using the techniques explained generate as many alternative concepts for solving Exercises 3: can opener, 4: vaciron, 6: multi-gym, 7: rail-cutter and 8: bilge filter which were outlined at the end of Chapter 2.

Concept selection

This chapter illustrates decision-making support methods for design concept selection using two case studies, that of fixing a gear to a shaft and the seat suspension mechanism introduced earlier. The argument that decisions can be made subjectively is refuted and the more popular formal concept selection methods are presented. These include the use of a decision tree, the datum method and design evaluation using a Harris diagram. A two stage approach is recommended in which the criteria from the specification are ranked and weighted and then, following design concept generation, each concept is evaluated against these criteria. Finally, the advantages of using a spreadsheet package on a personal computer coupled with the recommended method are highlighted.

4.1 Introduction

In Chapters 2 and 3 the PDS definition and the concept generation stages of the engineering design process were covered. The concept generation stage may be considered as divergent, in that the boundaries set in the PDS are fully explored. The next stage in the engineering design process involves selecting concepts and is convergent. It involves making decisions and rejecting the majority of concepts generated. The concept selection stage is clearly identified in Fig. 1.5 by means of the decision arrow between concept and detail.

As an illustration of the importance which must be attached to concept selection consider the two Figs 4.1 and 4.2. They show the cost and opportunity for change in product development and a representation of the all too frequent pattern of engineering changes. The indication of Fig. 4.1 is that during the very early stages of design there is

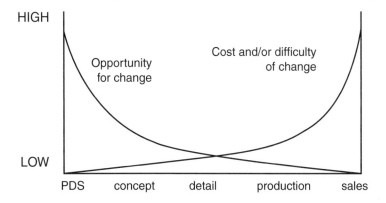

Figure 4.1 Cost and opportunity for change in product design

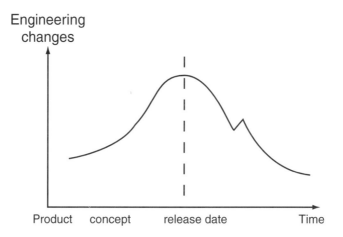

Figure 4.2 Typical development cycle indicating design changes

plenty of opportunity for making changes and suggesting improvements to the concepts put forward. However, once detail design commences this opportunity is substantially reduced and the cost of making changes increases exponentially with time as more resources are committed.

Figure 4.2 is perhaps more illuminating since it shows that the peak in design changes throughout the western world occurs much later than at the detail design stage and occurs around the release date. These changes are obviously very costly to implement and have a significant negative impact on profitability. The aim has to be to change the shape of the curve in Fig. 4.2 so that the peak occurs much earlier in the design process. The selection of an optimum concept which most nearly satisfies the PDS goes a long way to achieving this aim.

It is worth repeating that product cost and quality are largely pre-determined by design and that, as suggested in Chapter 3, practically all of product cost and quality are committed by decisions taken during the early design stages. It is generally acknowledged therefore that once a concept is selected the application of the most sophisticated design evaluation techniques cannot have a significant impact on product cost and quality.

The act of designing involves the designer in both intuitive and rational judgements and many decisions, on various levels, have to be made. Making decisions is stressful for most people and engineering designers are no exception to this rule. Reputation and self-esteem as a competent decision maker are at stake and the consequences for company viability of wrong decisions weigh heavy. However, the quality of a decision does not depend on the particulars of the situation but rather on the decision-making process adopted. The requirement, then, is a comprehensive decision support system for the designer.

An analogy with the decision-making process can be drawn from the manufacturing process. During manufacture raw materials are converted into products by certain processes. The quality of the final product depends equally on the quality of the raw materials and the quality of the processing. Both the product and the raw materials are visible. Any defect found can be traced to the raw materials or to the process. Fault finding is straightforward since the processing equipment is visible. One of our goals is to make the decision-making process of engineering design just as visible so that decisions made can be evaluated effectively.

There are many reasons why a formal decision-making process should be used by engineering designers:

(1) Time wasted in pursuing wrong alternatives to the detail design stage is avoided.
(2) Causing decision-making to be visible helps ensure the process is repeatable.
(3) The ability to evaluate the thought processes of others is developed.
(4) The designer can defend decisions made in discussions with managers or clients.
(5) A designer with no previous experience can carry out a sensible evaluation of alternative concepts.
(6) The process of concept selection stimulates new concepts or encourages combination of concepts.

It is at this stage in the design process that the first design audit meeting is convened, see Chapter 8. The design team must present both the PDS and the concepts generated to a gathering of colleagues from many other departments and justify the selection of the concept(s) which will be developed further.

As stated in Chapter 2 the writing of a realistic and comprehensive PDS is a fundamental part of any design. It is especially important since most formal methods of concept selection assess how well each concept fits the technical specification. At the very least, each concept is judged on the basis of company objectives along with some objectives derived from the specification.

The aim must be to reduce the number of concepts generated to a shortlist of not more than two or three for future detailed analysis. It is clearly impractical to examine every original concept in depth since resources of time, money and workforce are limited. If only one or two requirements are to be met the problem of identifying the most promising concept can be straightforward. However, if many objectives or criteria have to be met the situation can become very confused and a systematic evaluation method is essential.

Consider the selection of a method for fastening a gear to a shaft. It is possible to generate many varied concepts for even this relatively simple task. The sketches in Fig. 4.3 show only six common ways.

- Concept 1 involves *pinning* the gear to the shaft and the illustration shows a pin perpendicular to the axis of the shaft. The pin perpendicular to and through the centre of the shaft is to be preferred to a pin parallel to the axis if axial space allows since the pin will be in double shear.

- Concept 2 shows a *rectangular key*. There are many types of key, some of which do not require separate axial fixings. In the illustration a circlip is used for axial positioning.

- Concept 3 shows the centre of the gear extended and slit. This allows the *clamp* shown in end view to be tightened around the extended portion, using the integral screw, thus clamping the gear to the shaft.

- Concept 4 involves machining both the hole in the gear and the larger diameter of the shaft to fine tolerances, to ensure a *press fit*. A variation on this is to heat the gear, enlarging the hole, and cool the shaft to allow assembly by *shrink fit*. Such a design allows the transmission of much greater torques than with a press fit.

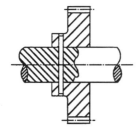

Concept 1 – Pinning

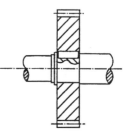

Concept 2 – Rectangular key

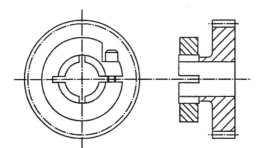

Concept 3 – Clamping

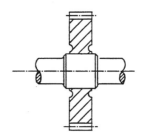

Concept 4 – Press or shrink fit

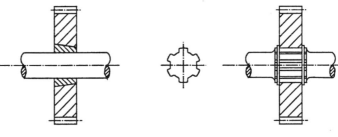

Concept 5 – Tapered bush

Concept 6 – Spline

Figure 4.3 Concepts for fixing a gear to a shaft

• Concept 5 is a *tapered bush*. The central bush and the inside diameter of the gear have tapers and the two are forced together, usually by means of grub screws, thus clamping the assembly to the shaft.

• Concept 6 can be considered as employing multiple keys as illustrated in end view. As with keys a *spline* needs to be fixed axially and in the illustration two circlips are used.

How does the designer decide which concept best satisfies the design brief, and more particularly, the PDS? The bulk of this chapter is devoted to a discussion of various design decision-making methods illustrated using the gear-fixing example. In the final section the recommended combination of methods is applied to the on-going example of the seat suspension unit.

4.2 Subjective decision-making

Table 4.1 illustrates some of the considerations that the design team must take into account when selecting the optimum alternative, in this case for the example of fixing a gear to a shaft. In fact the table could have been a page taken from an engineering designers note-book. The table has concepts on the horizontal axis and objectives on the vertical axis. The designer fills in each box of the table with a subjective assessment of how well each individual concept meets each individual objective. For example, concept 2, keying, gives excellent torque transmission but causes high stress concentrations.

Table 4.1 Subjective selection matrix for gear fixing

Objective	Concept					
	Pinning	Keying	Clamping	Press fit	Taper bush	Splining
Torque capacity	excellent	excellent	good	fair	excellent	excellent
Ease of gear replacement	poor	excellent	excellent	fair	excellent	excellent
Reliability in operation	excellent	excellent	fair	good	excellent	excellent
Satisfy environmental specification	excellent	excellent	good	good	good	excellent
Machining requirement	high	high	moderate	moderate	moderate	high
Ability to use pre-hardened parts	poor	excellent	excellent	excellent	excellent	excellent
Stress concentration effects	high	high	moderate	moderate	moderate	high
Relative cost	high	high	medium	medium	high	high
Axial location	yes	no	yes	yes	yes	no

Remarks A Pin can be designed to shear if gear is overloaded.
Press fit is suitable where the shaft is too small for splining.
Splining is ideal where the gear must slide along the shaft.
Tapered bushes are restricted to shaft diameters greater than 12 mm.

Notice that the list of concepts and objectives is not comprehensive, because it has been simplified for illustrative purposes. Many other objectives, such as assembly difficulty and balancing of gear and shaft, could have been included.

A very experienced designer rarely resorts to putting thoughts down, preferring instead to weigh all factors and to select a concept by intuition, without the need to resort to a formal procedure. However, this tendency is an error. As is illustrated in Table 4.1, even a relatively simple problem like the selection of a method for fastening a gear to a shaft can contain many underlying factors which are difficult to weigh against each other during a subjective assessment. Thus the adoption of a systematic decision-making procedure is essential for student engineers and the most experienced designers alike during concept selection.

Decision-making in design is perhaps best likened to the application of a series of filters, some coarse, some fine, which reduce the original list of concepts to manageable proportions. The coarse filters are applied initially and can include basic calculations to establish non-viability, experimental work of a simple nature and simple modelling. Table 4.1 could be considered a coarse filter, since it would still leave the final selection of the concept as subjective and open to individual bias. Nevertheless, any form of documentation of the decision-making process is better than none at all and a more reasoned selection of the best concept would result from adopting such an approach.

A more rigorous application of decision-making techniques than that described has been proven to lead to better overall solutions to design problems. However, it is not suggested that a designer should always employ any of the following decision-making processes in isolation. On the contrary, best results are achieved when a team comprising the designer and other company personnel is formed.

Design engineers are often of the biased opinion that they are the only people capable of making sensible decisions. This is patently not the case and company employees from many other sections make just as many, if not more, decisions as designers in the course of their normal duties. Also, when working on a complex project it is always wise to involve the client in the decision-making.

4.3 Criteria ranking

The first step in formal concept selection procedures is to rank the PDS criteria in order of relative importance. Most methods of criteria ranking involve the formation of a matrix and one method, proposed by Pugh, is perhaps the most widely used. This matrix has become known as the binary dominance matrix.

In this method a matrix is formed, similar to that shown in Table 4.2 for the gear-fixing example, with the criteria listed on both the vertical and horizontal axes. A 1 or 0 is placed in each box of the matrix depending upon the relative importance of the pair of criteria. It is suggested by some that for criteria judged of equal importance a value of 0.5 be allocated to each. However, the view taken by the majority of those with experience of using such methods is that these relatively minor decisions must be forced, otherwise it is all too tempting to take the easy way out and allocate many values of 0.5, thus rendering the whole process almost worthless.

Table 4.2 Criteria ranking matrix

Criteria	A	B	C	D	E	F	G	H	I	Total	Rank
A Torque capacity	\	1	1	1	1	1	1	1	1	8	1
B Ease of gear replacement	0	\	0	1	1	1	0	0	1	4	5
C Reliability in operation	0	1	\	1	1	1	0	0	1	5	4
D Satisfy environmental spec.	0	0	0	\	0	0	0	0	1	1	8
E Machining requirement	0	0	0	1	\	1	0	0	1	3	6
F Ability to use pre-hardened parts	0	0	0	1	0	\	0	0	1	2	7
G Stress concentration effects	0	1	1	1	1	1	\	0	1	6	3
H Relative cost	0	1	1	1	1	1	1	\	1	7	2
I Axial location	0	0	0	0	0	0	0	0	\	0	9
										36	

The procedure used and illustrated in Table 4.2 is to consider each row in turn. In the first instance decisions are made regarding whether torque capacity is more (or less) important than all other criteria in turn. If the criterion on the vertical axis (torque capacity) is considered more important than that with which it is being compared on the horizontal axis, a value of 1 is allocated. Hence values of 1 are allocated for the first row, torque capacity being judged more important than all other criteria. Each comparison of a pair of criteria is made twice, above and below the diagonal. The decisions made for the second time form a useful check for consistency and are necessary for completeness of the number of 1s allocated to each criterion.

In a large criteria ranking matrix a useful check that all decisions have been made and that any decision has not been contradicted when the pairs of criteria are considered against each other for the second time is that the total number of 1s must equal $0.5n(n-1)$, where n is the number of criteria. In this case 9 criteria were used, therefore 36 1s were allocated, as is verified by adding the totals column.

When all pairs of comparisons have been made, the row totals indicate the positional importance, or rank order, of the criteria. The next stage in the decision-making process is to assign values representing the relative worth of each criterion.

4.4 Criteria weighting

During the weighting procedure the criteria are first re-ordered, with the most important first, as in Table 4.3. This re-ordering ensures that those criteria which have the greatest influence on concept selection are considered first. This is particularly important when many criteria are involved and many decisions have to be made. The assigning of weightings to each criterion is the subject of much debate. There are even those who suggest that rank-ordering is an example of an ordinal scale and that arithmetical operations cannot be performed on an ordinal scale. Strictly speaking, this is true, but the aim here has to be to assist the designer in any way possible. Without a formal method of applying weightings the designer is again left with the stressful business of subjective weighting allocation and therefore subjective decision-making.

Table 4.3 Criteria ranking and weighting matrix

Criteria	Total	Rank	Weighting
A Torque capacity	8	1	0.222
H Relative cost	7	2	0.194
G Stress concentration effects	6	3	0.167
C Reliability in operation	5	4	0.139
B Ease of gear replacement	4	5	0.111
E Machining requirement	3	6	0.083
F Ability to use pre-hardened parts	2	7	0.056
D Satisfy environmental spec.	1	8	0.028
I Axial location	0	9	0.000
	36		1.000

One rigorous method of assigning weightings is to use a decision tree. The decision tree is a device for presenting the structure of a decision in an objective manner and although the technique is more commonly applied in project management, where decisions are made in succession into the future, it has some merit for design criteria weighting. The decision tree is a means of integrating all the factors and probabilities that make up the problem, and presenting them in a quantified form. The adoption of such a technique has the advantage, in common with other methods, that it at least prevents the biased proposals of the loudest talker being forced through.

At the highest level the overall objective is given a value of 1.0. At each lower level of the tree the sub-objectives are given weights relative to each other, but which total 1.0. Their true weights are then calculated as a fraction of the weight of the objective above. Using this procedure it is easier to assign weights with some consistency because it is relatively easy to compare sub-objectives in small groups of two or three and with respect to a single higher level objective. Also, all the true weights add up to 1.0 ensuring arithmetic validity of the weights.

The objectives weighting tree for fixing a gear to a shaft is shown in Fig. 4.4. It should be noted that the specification criteria used in earlier methods had to be modified into objectives and sub-objectives. In this instance four levels of objectives were identified, level 1 being the overall objective. At level 2 the three main objectives were identified as torque capacity, production or ex-works cost and reliability. Levels 3 and 4 contain the sub-objectives on which the main objectives depend.

The weightings assigned to individual objectives are displayed on the right side of the figure and of course their sum is 1.000. The allocation of the first value in each box represents the designers subjective assessment of the relative importance of those objectives being compared. For example, in level 2; ex-works cost, torque capacity and reliability are allocated values of relative importance of 0.3, 0.4 and 0.3 respectively. At level 3, ex-works cost is divided into three sub-objectives; simple assembly, use of bought out parts and simple machining. The first figure in each of these boxes is the relative value, again allocated subjectively, which the designer feels each warrants. However, since ex-works cost as a whole can only total 0.3, each of these figures, 0.3, 0.2 and 0.5, must be multiplied by 0.3.

The second figures in each box are the relative weightings of each objective. Consider the weighting factor for torque capacity. It is 0.4 since this cannot be broken down into sub-objectives. For ease of gear replacement the factor is significantly less at 0.036 since this is a sub-objective of ease of maintenance which in turn is a sub-objective of reliability. Weightings are allocated from left to right with the second figures in each cluster of sub-objectives totalling the second figure of their higher level objective.

The method of weightings allocation recommended for use is presented in Table 4.2. The criteria are allocated a normalized weighting factor by dividing the total number of decisions made (for the gear-fixing example this is 36) into the individual totals. Obviously the sum of all the weighting factors must be 1.0, as shown. Also, it should be noticed that in Table 4.3 the criteria have been re-ordered with the most important at the top.

As stated earlier most concept selection methods rank and weight the criteria or objectives as a first step. However, this chapter would not be complete without mention of other methods of concept selection in which no attempt at ranking or weighting is made. Two such methods, the Datum Method, first proposed by Pugh, and EVAD (Design EVAluation) are worthy of detailed consideration.

4.5 Datum Method

The Datum Method is based on the use of a matrix having concepts on one axis and criteria (objectives) on the other, the criteria being a basic form of the PDS. One concept is selected

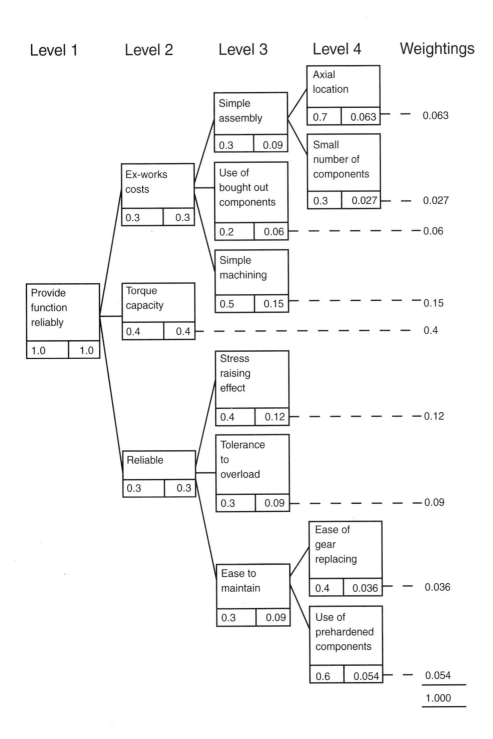

Level 1	Level 2	Level 3	Level 4	Weightings

Figure 4.4 Objectives tree for fixing a gear to a shaft

as a datum and each other concept is considered in comparison with this for each criterion. The categories are 'better than' (+), 'same as' (s) and 'worse than' (–). After this first iteration one concept should emerge as best fitting the criteria.

Again, consider the example of fixing a gear to a shaft. For the first iteration shown in Table 4.4, the purely arbitrary choice of concept 1, pinning, as the datum was made. The other columns contain the comparative decisions made during this first run through. Each other concept was compared against the datum and each criterion. For example, the allocation of a (+) under splining for torque capacity indicates that in the designer's opinion more torque can be transmitted using a spline than by using the datum, pinning.

Table 4.4 Datum method of concept evaluation – first datum

Objective	Concept					
	1 Pinning	2 Keying	3 Clamping	4 Press fit	5 Taper bush	6 Splining
Torque capacity	First datum	s	–	–	+	+
Ease of replacement		+	+	s	+	+
Reliable operation		s	–	–	s	s
Satisfy environment		s	–	–	–	s
Machining requirement		–	+	+	+	–
Pre-hardened parts		+	+	+	+	+
Stress concentrations		s	+	+	+	+
Relative cost		s	+	+	–	–
Axial location		–	s	s	s	–

Once all the individual decisions have been made the numbers of pluses and minuses are totalled for each concept. In direct comparison with pinning, concept 5, taper bushes, proved marginally better than concept 3, clamping, scoring 5+, 2– and 2s.

For subsequent iterations the strongest concept is made the datum and evaluated against each of the other concepts. In practice it is usually found necessary to carry out three iterations, as Table 4.5 illustrates. In the second iteration the concept with the highest positive balance, concept 5, was chosen as the datum and the columns headed 'b' contain the decisions made. As a result of this second run through, concept 6, splining, with more pluses than minuses was chosen as the third and final datum.

On the basis of the decisions contained in the columns headed 'c' concept 3, clamping, was chosen as the best overall concept scoring better than all three chosen datums, although the difference between taper bush and clamping was extremely marginal and in practice, if resources of time and money allow, both would be developed further.

In common with most decision-making techniques, the Datum Method involves making many minor, but no major, decisions. Such comparisons avoid the need for making absolute measurements and by recording each decision in matrix form the need to memorize the decisions made is avoided. It is the sheer number of decisions that ensures a

Table 4.5 Datum method of concept evaluation – complete chart

Objective	Concept 1 Pinning			2 Keying			3 Clamping			4 Press fit			5 Taper bush			6 Splining		
	a	b	c	a	b	c	a	b	c	a	b	c	a	b	c	a	b	c
Torque capacity	F	–	–	s	s	–	–	s	–	–	–	–	+	S	–	+	+	T
	i													e				h
Ease of replacement	r	–	–	+	+	s	+	s	s	s	–	–	+	c	–	+	+	i
	s													o				r
Reliable operation	t	s	s	s	s	–	–	s	–	–	–	–	s	n	–	s	+	d
														d				
Satisfy environment	d	+	s	s	+	s	–	s	s	–	s	s	–	d	–	s	+	d
	a																	a
Machining requirement	t	–	+	–	–	+	+	s	+	+	–	+	+	a	+	–	–	t
	u													t				u
Pre-hardened parts	m	–	–	+	s	s	+	s	s	+	s	s	+	u		+	s	m
														m				
Stress concentrations		–	–	s	–	–	+	s	+	+	s	+	+		+	+	–	
Relative cost		+	+	s	+	+	+	+	+	+	+	+	–		s	–	s	
Axial location		s	+	–	–	s	s	s	+	s	s	s	s			–	–	

final outcome which should not vary from one designer, or design team, to another. Although it is probable that any student or design engineer considering the gear problem would disagree with some of the individual decisions made, the overall result of the concept selection process should always be the same when identical criteria are used.

The three major drawbacks with the Datum Method of concept selection are that all decisions are made several times (three in this case), making the process time consuming; that no account is taken of the relative importance of the criteria applied; and that no scale is used to indicate how much each concept is better or worse than the datum.

4.6 EVAD (Design EVAluation) method

The EVAD method of concept selection was developed at the University of Twente. As with the Datum Method no attempt is made at ranking or weighting the criteria. It is a method particularly recommended for the evaluation and selection of new product ideas but it can also be applied to concept selection. The stated aim of the method is to support the designer during concept selection whilst still retaining an intuitive aspect.

In common with all other methods it is first necessary to reduce the usual long list of generated concepts to manageable proportions. It is recommended that a maximum of six concepts should be chosen for further investigation and that a few major criteria should be chosen from the specification to assist this process. For example, when reviewing new products this initial selection could be made subjectively by considering the following questions:

- What is the probability of market success?
- What will be the magnitude of business generated?
- Is the idea compatible with the company's capabilities?
- What is the likely return on investment?

Table 4.6

Criterion	Standard	Score
Torque capacity	High	++
	Moderate	+
	Low	−
	Non-existent	−−
Ease of gear replacement	By hand	++
	With tools	+
	With special tools	−
	Not possible	−−
Reliability in operation	Infinite life	++
	10^6 cycles	+
	10^6 cycles with servicing	−
	Under 10^5 cycles	−−
Satisfy environmental specification	Completely robust	++
	Affected by hostile env.	+
	Affected by most env.	−
	Unsatisfactory	++
Machining requirement	None	++
	Little	+
	General machining	−
	Specialist machining	−−
Ability to use pre-hardened parts	Yes	++
	With modification	+
	With difficulty	−
	No	++
Stress concentration effects	None	++
	$K_f < 1.2$	+
	$1.2 > K_f < 1.6$	−
	$K_f > 1.6$	−−
Relative cost	Low	++
	Medium	+
	High	−
	Substantial	−−
Axial location	Yes	++
	Yes with modification	+
	Yes with difficulty	−
	No	−−

The six selected concepts are subjected to a formal evaluation procedure, according to a modified form of what is known as the Harris method. In this method a list of evaluation criteria is established and each criterion is standardized according to the strategic objectives of the company. For example, one criterion, product image, could have standards set as modern exclusive (++), modern ordinary (+), traditional (−) and outdated (−−). From the concept descriptions scores for each criterion are established. These results are entered on a diagram giving an evaluation profile for each idea. The profiles are compared in a qualitative rather than quantitative way.

Although EVAD is particularly recommended as a method for selecting new product ideas it can be equally well applied to any part of the engineering process. Consider again the illustrative example of fixing a gear to a shaft. An EVAD evaluation of this, assuming the initial list of concepts had already been reduced to six would start by standardizing the criteria contained in the specification. Standardization takes the form of establishing the complete range for each criterion and dividing this equally into four sub-divisions, as shown in Table 4.6.

Once the criteria have been standardized a diagram is produced for each concept. As in the Datum Method the criteria form the vertical axis and concepts the horizontal axis. Four columns are necessary for each concept ranging from (−−) to (++) and using standards such as those established for the gear fixing a decision is made for each concept against each criterion. Table 4.7 contains the diagrams for concepts 3, 5 and 6, clamping, taper bush and splining, from the gear-fixing example. In practice all six concept diagrams would have been plotted.

Table 4.7 Harris diagrams for clamping, taper bush and splining

Concept evaluation sheet

| | | *Ideas* | | | | | | | | | | |
| | *3 clamping* | | | | *5 taper bush* | | | | *6 splining* | | | |
Criteria	−−	−	+	++	−−	−	+	++	−−	−	+	++
Torque capacity			X				X					X
Ease of gear replacement			X			X					X	
Reliability in operation			X				X					X
Satisfy environmental spec.			X				X				X	
Machining requirement			X				X		X			
Ability to use pre-hardened parts			X					X				X
Stress concentration effects			X					X	X			
Relative cost			X		X				X			
Axial location			X					X			X	

Remarks

In our example it is clear that clamping has the most positive profile. It would again be selected, as it was by using the Datum Method. However, as stated above, the aim in drawing the Harris diagram is to present the designer with a visual representation of the decisions to be made and not to provide a quantitative assessment of the alternative concepts. The designer is only guided in selecting a concept and must still use intuition and experience whilst remaining aware that the criteria are not weighted and that the scoring is subjective.

Obviously this is a relatively simple example of the EVAD method involving only a few criteria. In a more complex problem a single concept would be represented by a diagram covering a complete sheet of paper, the Xs replaced by boxes colour-coded in four different colours depending upon the score allocated and remarks would be made explaining the thinking behind some of the less straightforward decisions. All six diagrams would be compared visually and those concepts with a more positive profile would be regarded as worthy of further investigation.

4.7 Recommended concept selection method

In Chapter 2 the PDS was defined and in Chapter 3 concepts were generated for the example of the seat suspension mechanism. Here the concept with the highest potential is selected using those techniques most suitable for decision-making by an inexperienced engineer.

First, as shown in Table 4.8, the functions and constraints are ranked in order of relative importance using the matrix method. In this case there are 20 criteria to be ranked and using the checking formula $0.5n(n - 1)$, 190 individual decisions are made. The criteria are then allocated a normalized weighting factor by dividing 190 into the individual totals.

Table 4.8 Binary dominance matrix for seat suspension mechanism

Criteria	1	2	3	4	5	6	7	8	9	10	11	12	13	14	15	16	17	18	19	20	Total	Weighting
1 Rotation	0	1	1	1	1	0	1	0	0	0	1	1	1	1	1	1	0	0	1	0	12	0.063
2 Adjustment <25 mm	0	0	1	0	1	0	1	0	0	0	1	1	0	0	1	1	0	0	0	0	7	0.037
3 Adjustment speed	0	0	0	0	1	1	0	0	0	0	0	0	0	0	1	1	0	0	1	0	5	0.026
4 Size restriction	0	1	1	0	1	1	1	0	0	0	0	0	1	1	1	0	0	0	1	0	9	0.047
5 Minimum weight	0	0	0	0	0	0	0	0	0	0	0	0	0	0	1	1	0	0	0	0	2	0.011
6 Fit all seats	1	1	0	0	1	0	1	0	0	0	1	1	1	1	0	0	0	1	1	0	10	0.053
7 Anchor seat belt	0	0	1	0	1	0	0	0	0	0	0	0	1	1	1	1	0	0	0	0	6	0.032
8 Fail safe	1	1	1	1	1	1	1	0	0	0	1	1	1	1	1	1	1	1	1	1	17	0.089
9 Secure locking	1	1	1	1	1	1	1	1	0	1	1	1	1	1	1	1	1	1	1	1	19	0.100
10 Damping	1	1	1	1	1	1	1	1	0	0	1	1	1	1	1	1	0	0	1	0	15	0.079
11 Straight travel	0	0	1	1	1	1	1	0	0	0	0	0	0	0	1	1	0	0	0	0	7	0.037
12 Ergonomics	0	0	1	1	1	0	1	0	0	0	1	0	0	0	1	1	0	0	1	1	9	0.047
13 Dirty environment	0	1	1	0	1	0	0	0	0	0	1	1	0	1	1	1	0	0	0	0	8	0.042
14 Open environment	0	1	1	0	1	0	0	0	0	0	1	1	0	0	1	1	0	0	0	0	7	0.037
15 Aesthetics	0	0	0	0	0	0	0	0	0	0	0	0	0	0	0	1	0	0	0	0	1	0.005
16 Quality	0	0	0	1	0	1	0	0	0	0	0	0	0	0	0	0	0	0	0	0	2	0.011
17 Manufacturing cost	1	1	1	1	1	1	1	1	0	0	1	1	1	1	1	1	1	0	1	1	17	0.089
18 Material cost	1	1	1	1	1	1	1	1	0	0	1	1	1	1	1	1	1	0	0	1	16	0.084
19 Maintenance	0	1	0	0	1	0	1	0	0	0	1	0	1	1	1	1	0	0	0	0	8	0.042
20 Reliability	1	1	1	1	1	0	1	0	0	1	1	0	1	1	1	1	0	0	1	0	13	0.068
																					190	1.000

The design team is now in a position to make an objective assessment of how well each concept generated during the creativity stage satisfies the PDS. The criteria are re-ordered with those allocated the highest weightings at the top. Again a matrix table is used in Table 4.9, but this time the criteria on the vertical axis are compared with each potential solution, see Fig. 3.11 for concept sketches of the seat mechanism. Each concept is scored as to how well it fulfils each criterion and allocated a percentage. If a zero is allocated to any concept then it does not comply with that particular criterion and therefore cannot satisfy the PDS as a whole. That concept is rejected without further work.

Once the percentages have all been allocated, the values are multiplied by the weighting factor for that particular criterion. These figures are added to give a total percentage satisfaction figure for each concept. Obviously, were any concept to fit the specification

Table 4.9 Complete concept selection worksheet for seat suspension

Criteria		Weighting	A%	B%	C%	D%	E%	F%
					Concepts			
9	Secure locking	0.100	80	70	80	50	60	60
17	Manufacturing cost	0.089	60	50	60	50	50	50
8	Fail safe	0.089	30	20	60	20	50	40
18	Material cost	0.084	60	80	60	70	60	60
10	Damping	0.079	90	60	50	50	40	60
20	Reliability	0.068	70	80	80	60	50	70
1	Rotation	0.063	100	100	100	100	100	80
6	Fit all seats	0.053	90	90	80	80	90	90
4	Size restriction	0.047	80	90	80	60	50	70
12	Ergonomics	0.047	60	50	50	60	70	60
19	Maintenance	0.042	80	80	70	60	50	70
13	Dirty environment	0.042	60	70	80	50	60	50
11	Straight travel	0.037	20	10	90	10	20	40
14	Open environment	0.037	70	80	80	70	60	70
2	Adjustment <25 mm	0.037	100	90	90	90	100	90
7	Anchor seat belt	0.032	90	60	70	90	60	70
3	Adjustment speed	0.026	50	60	60	10	50	60
16	Quality	0.011	40	70	80	50	40	50
5	Minimum weight	0.011	40	60	70	50	50	50
15	Aesthetics	0.005	90	70	80	90	70	70
			69.1	66.1	71.6	56.9	59.1	62.2

THE BEST CONCEPT SCORED 71.6

completely it would obtain a score of 100%. In this way, a measure is obtained of how well each concept satisfies the PDS along with a comparison between concepts.

Obtaining an absolute measure for specification satisfaction is important since if no concept scores a reasonable percentage the whole project should be reviewed and perhaps shelved. It is not possible to quantify a minimum percentage which is acceptable since all design problems are different in nature and individual designers will be more (or less) optimistic in the individual values allocated. In the seat suspension example, concepts C and A, in that order, have demonstrated that they are worthy of further investigation.

The choice of concept evaluation technique must eventually rest with the individual designer or be dictated by company policy. All of the processes mentioned have their merits and it would be wrong to pretend that any one produces superior results than another. However, the process used to select a concept for the seat suspension mechanism probably represents the decision-making process best employed.

4.8 Computer aided decision making

All of the techniques described are time consuming and a little tedious in operation. One way of improving the situation would be to computerize the process. A type of standard computer program which could be employed with the recommended concept selection method is a spreadsheet. These programs are widely available on personal computers and are commonly employed in financial analysis. There are many advantages in their use in the concept evaluation process and the method used for the seat suspension mechanism concept selection could readily be input to a spreadsheet.

In the use of a spreadsheet for criteria ranking the totals and normalized weighting factors would be computed automatically using previously defined formulae. Using the program's sorting capability the criteria could be re-ordered. There would be no need for a separate concept evaluation matrix which could be combined with the criteria ranking and weighting matrix. Once all scores were allocated by the designer for how well each concept satisfies each criterion the computer would indicate the totals and highlight the highest score.

Having set out the format of a worksheet any subsequent project would be performed simply by editing that worksheet. Some criteria will be common to all designs, others will be different. A spreadsheet could act as a limited form of expert system if common criteria and common decisions, based perhaps on company policy, were stored and re-used. The designer would have fewer decisions to make, leaving more time for genuine creativity.

When compared with paper-based methods, spreadsheets accelerate the selection procedure. Rapid sensitivity analysis on the influence of individual decisions to the overall result is possible, as is rapid consideration of the effect of minor design changes to a concept. One final point is that an identification of weak spots where the chosen concept does not score highly against particular criteria or objectives can often lead to improvements in the design.

4.9 Principles

Concept selection principles

Convergence During concept generation the problem is broadened. During concept selection the process of convergence towards a single solution is begun.

Intuition Concept selection methods vary from the purely intuitive, which is to be avoided for complex problems, to the formal and rigid. However, there is always room for some intuition and professional judgement in the evaluation and decision-making process.

Visibility Decisions made must be transparent and easily recorded.

Ranking The relative importance of the design criteria can be augmented by the addition of relative weightings.

Comparison When the formal structure for decision making is followed a numerical means of comparison between projects is provided. The score for the best concept gives supporting information as to the likely prospects for a successful, profit-making, design.

4.10 Exercises

1. Consider the wide variety of devices available for assisting the removal of a cork from the neck of a wine bottle. Select the optimum using the recommended decision-making process.

2. Identify the concept which best meets the specifications you have written for the rail-cutter, the bilge filter problem, and the one-handed can opener introduced in the exercises at the end of Chapter 2.

3. A survey has been conducted which indicates a sizeable market for a power-driven mini-trencher. The use would be mainly for services on new housing estates, where a trench 100 mm wide by 450 mm deep is the minimum requirement, and for digging drainage channels in lawns. It is envisaged that sales would be almost solely through Contractors/Dealers who would then hire the device out. Write a PDS, generate concepts and select the optimum concept for the mini-trencher.

5 Embodiment

A more detailed analysis of the selected concept(s) is undertaken in the embodiment stage of the design process. Subjects covered include form design, design for manufacture and assembly, materials and process selection and industrial design. However, the main aim of this chapter is to establish concept development as a distinct stage in the design process by identifying the steps and rules employed. Materials and process details are not included since this information is obtainable elsewhere. Design is not solely the achieving of technical solutions but also creating useful products which satisfy and appeal to their users. There are three broad areas of design activities, technical, ergonomic and aesthetic. The overlap area of ergonomics between the engineering and industrial designer is also covered.

5.1 Introduction

An essential element of the continual cycle of deliberation and verification of the design proposal is that it must be checked periodically to establish that it is indeed optimum. Once a scheme takes shape efforts should concentrate on optimization of the design. This is usually accomplished by 'attacking' what are perceived to be the weak points of the design.

The embodiment process is the bridge between the conceptual stage of the design process and the detail design stage. The aim is to refine and develop concept sketches, such as those presented in Fig. 3.11, to such an extent that detail design and production planning can commence. The input to the embodiment stage is often little more than an outline sketch and associated project controlling documentation such as PDS or design requirements. The output is a definitive scheme drawing accompanied by documentation, such as calculations, required tolerances and suggested materials and manufacturing processes, which explains the design intent.

Concept C, which was the selected concept for the seat suspension exercise, is shown in Fig. 5.1 to illustrate these aims. It must be remembered that although a 3D representation is very useful during the embodiment stage it is of no use during the detail stage and scheme drawings, drawn strictly to scale, are essential.

The isometric drawing in Fig. 5.1 highlights many embodiment issues. These issues are many and the more obvious include:

- How is the suspension travel to be provided?
- What is the required size and strength of structural components?
- Can a damper of the required proportions be purchased?
- Are adjustment handles ergonomically acceptable?
- Will welded joints be acceptable?
- What are the implications for manufacture and assembly?

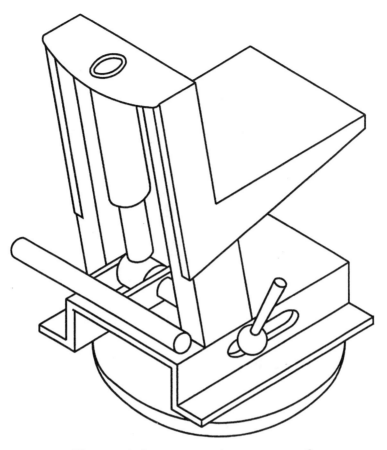

Figure 5.1 Seat suspension – concept C

The inevitable conclusion from such thinking is that the final design will involve compromise between conflicting requirements. Many actions must be performed at the same time and the solution synthesized since decisions made in one area can have a knock on effect elsewhere.

The embodiment process is illustrated in Fig. 5.2. It is cyclical or iterative in nature following broadly the pattern indicated in the outer ring of the figure. The process begins with a decision being made on overall layout. This is then modelled, analysed, synthesized and optimized. A revised layout is then produced or more detail added to the original layout and the embodied design is evaluated against the functions and constraints in the PDS. The whole process is repeated many times and for different areas of a design, until the best compromise solution, taking into account manufacturing and assembly processes and materials selection, is reached.

The value engineering process runs in parallel with the embodiment process. Value engineering establishes the cost and performance of alternative proposals and provides useful information to the design team. Synthesis of all the information available at the embodiment stage, including cost estimates, guides the team in the development of an optimum design proposal. The following sections outline in more detail the embodiment process.

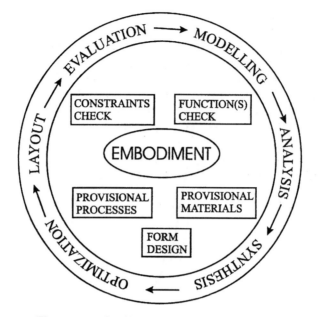

Figure 5.2 Cyclic nature of embodiment

5.2 Size and strength

Identify those functions and constraints within the PDS which will influence and determine overall size and strength. Specified geometry restrictions are deterministic, such as 'the radius about the centre of rotation < 300 mm' which is in the seat suspension PDS and should be added to the scheme drawing. Those factors which influence size and strength include operator weight and size, maximum frequencies of vibration, required factors of safety, etc.

5.3 Scheme drawing

The scheme drawing should be started and all known parameters included. The scheme should include all motion, to ensure adequate clearances, and copious notes indicating decisions made on subjects such as tolerances and materials. The scheme drawing is regularly updated and modified as the embodiment process continues and as more decisions are made. This is one important method of controlling the knock on effects and implications of decisions. Since the scheme is drawn strictly to scale only very important dimensions, usually those which include tight tolerances, are included.

As the embodiment process continues more information can be added to the scheme. Standard and bought out parts can influence decisions greatly since sizes and availability are normally set and restricted respectively. In the seat suspension example the damper would be a bought out item. The required travel is dictated by the PDS and once the likely maximum weight of the operator is known a decision could be made as to the exact damper required. Once the availability of the damper has been verified then the geometry is added to the scheme.

5.4 Form design

Once some shape and body begins to emerge then the modelling, analysis and synthesis process can begin. In the early stages the required sizes and strengths of major components are established. As the process continues and more knowledge is gained then issues such as form design and stress flows in any joints are considered. This process can be further divided by considering the primary function first and then repeating the cycle for each secondary function. It is important that, during this phase, some provisional thought centres on the likely manufacturing and assembly processes.

There are many factors which must be considered in the early stages of defining the shape and form of components and structures. It would clearly be a mistake to design any component without regard to the manufacturing process which is to be used. Decisions made at the embodiment design stage dictate manufacturing processes and it is imperative that manufacturing engineers, if not part of the design team, are at least consulted before decisions are made. There are obviously advantages and disadvantages to each manufacturing process and often the decision is made on cost against quantity grounds.

Consider the design of a link which must have a central pivot and two guide holes at its extremities. The likely manufacturing processes to be considered are casting, drop forging, hammer forging and welding. Figure 5.3 indicates how the cross-section and overall shape

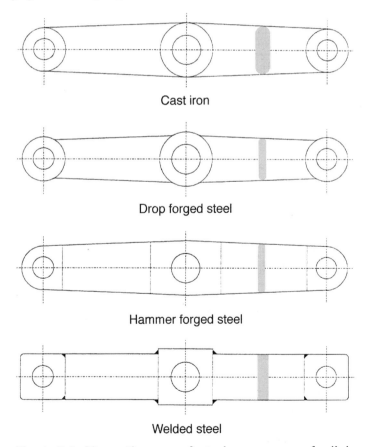

Cast iron

Drop forged steel

Hammer forged steel

Welded steel

Figure 5.3 Alternative manufacturing processes for link

would be influenced by each process. Cast iron is not as strong in tension and bending as steel so more material is required and the cross-section must be much thicker. Drop forged steel will use the least material of all. The decision as to which manufacturing process to choose is always significantly influenced by the quantities to be manufactured.

A graph of manufacturing cost against quantity is given in Fig. 5.4. Whilst it must be recognized that each component will have slightly differing characteristics the general trends remain the same. Clearly for large quantities drop forging using dies is the cheapest process and has the added bonus of giving the highest strength to weight ratio. For small quantities then a welded construction is probably best.

Load paths have a great influence on the shape and form of components. The ideal is often to try and design so that components are subjected to pure tension and compression.

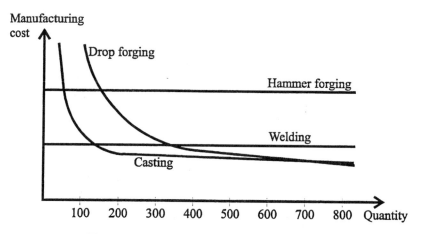

Figure 5.4 Influence of quantity on cost

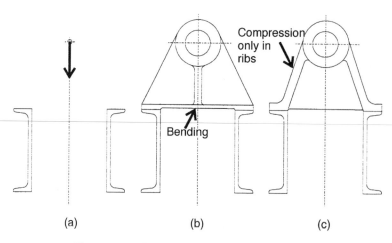

Figure 5.5 Smoothing load transmission

Consider the problem of mounting a bearing block onto two existing channel members, illustrated in Fig. 5.5. One of the prime requirements in the design of the block is to ensure smooth transition of load from the bearing into the two channel members. Figure 5.5(a) indicates the requirement, which is to transfer a compressive load from the bearing, through the block and into the flanges. A poor design is indicated in Fig. 5.5(b) since, with the vertical web the load will cause bending of the flange of the block directly under the centre of the bearing. In the suggested design in Fig. 5.5(c) the webs are in compression only and the stresses caused by the load are smoothly transferred through the bearing block.

This type of consideration can lead to the challenging of long held beliefs with surprising results. For example, one very common design of a component for translating reciprocating horizontal motion into reciprocating vertical motion is the bell-crank lever. However, as illustrated in Fig. 5.6(a), this design has to be relatively thick to prevent deflection due to bending. The alternative in Fig. 5.6(b), which at first sight looks wrong because we do not commonly see designs like it, can involve 50% less material because bending is almost completely designed out.

A further effect which influences form design is that of size. The effect of size can be illustrated, see Fig. 5.7, by considering again the bearing block. The larger the height dimension becomes the lighter or thinner the form should become. The centre section of the larger block is thus hollow.

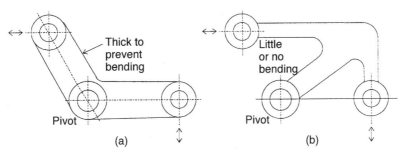

Figure 5.6 Modern 'bell-crank' lever

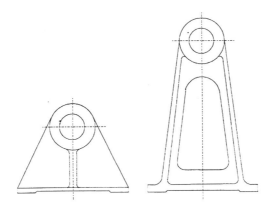

Figure 5.7 Effect of size on form

Rules

- Identify the optimum manufacturing method at the earliest possible stage taking account of quantities to be manufactured, strength requirements, weight restrictions and any other relevant factors.
- Place material so that it follows the same direction as the lines of force.
- Smooth the transfer ('flow') of stress due to load through components.
- Take careful note of the size of components and design shape and form to suit.

5.5 Provisional materials and process determination

Material and process selection is an integral part of the engineering design decision-making processes. It is essential that a materials and process audit is carried out as part of each design audit. Materials and processes are developing so quickly that it is difficult for a designer to have a thorough and up to date understanding of all modern materials. As an illustration of this consider the fact that there are well over 15 000 engineering materials to choose from. Modern materials and processes are having a significant effect on industry, increasing the options for the design and manufacture of new products. Materials account for approximately 50% of the cost of an average manufactured product. A further complication in the selection of materials and processes is the possibility of improving component characteristics by surface coating the component. For example, a titanium nitride coating on a high-speed tool drill enables it to drill ten times as many holes before re-sharpening and saves the cost of the coating many times over.

The proper use of materials leads to increased product performance, greater efficiency and reduced costs, resulting in increased competitiveness for companies. The substitution of plastic for metal components can lower assembly costs by reducing the number of parts and cut processing costs. Engineering ceramics continue to replace more traditional engineering materials in a wide range of applications. This is due mainly to their excellent corrosion resistance at both high and low temperatures. An example is silicon carbide which is being used for pump bearings, providing prolonged life and reducing the need for lubrication. Over the last five years the reliable strength of ceramics has doubled, plastics can be made fire resistant and cast metals can be processed to have the properties of forgings. At the limits of science and technology, materials are being developed for specific tasks. These are specialized materials with little general applicability. In such circumstances the material the designer must employ (and the associated process) in the solution would form one of the constraints in the specification.

If the specification is thorough and complete, the selection of material and process should be clearly constrained. From the specification, the main criteria may be identified and can often be a ratio such as cost/unit volume, cost/weight or strength/weight. Indeed, quantitative methods of selection are often based on ratios such as these. The decision-making process is best carried out as an elimination process. Start by excluding those materials and processes which clearly are not suitable.

The main criteria used to select the combination of materials and processes are:

- availability
- quantity required
- vibration damping
- cost
- tolerances required
- environmental impact

- density
- surface roughness
- ease of machining
- styling possibilities
- friction coefficients
- electrical properties

- wear resistance
- speed of delivery
- corrosion resistance
- operating environment
- chemical resistance
- mechanical properties.

Composite materials allow the designer to tailor materials and associated processes to meet the exact specifications of the product, but, if this choice is to be considered seriously the manufacturing process must be considered from the concept. It has been acknowledged for a long while that composite materials have a high strength to weight ratio, exceptional stiffness and excellent corrosion resistance. However, their use has been restricted due to high production costs. As alternative production methods to 'hand lay up', such as 'resin transfer moulding' and 'sheet moulding compound' reduce the production costs so the use of composite materials will extend beyond specialized components. As an example, composite leaf springs are now quite common since they are much lighter than steel ones, are inherently corrosion resistant, have a better fatigue life and a gradual rather than sudden failure mode.

When a new product is launched there are many uncertainties and it will be a brave designer who will experiment with a new or unfamiliar material or who will push the material to the limit of its performance. However, if opportunities to maximize competitive advantage are not to be missed the engineering designer's knowledge of advanced materials and processes must be kept up to date.

5.6 Design for assembly and manufacture

Any engineering designer, whether or not working as part of a team with manufacturing engineers, requires a working knowledge of manufacturing methods. Good practice dictates that during all stages of the design process advice is sought from manufacturing experts and the designer should attempt to utilize existing machinery and tooling where possible.

Assuming that possible interference of components or gross errors in the 'logic' of assembly are detected during the execution of the scheme drawing, a design should be suitable for machining and assembly if the following are critically appraised:

- ease of machining
- economy
- use of existing machinery and tooling
- avoidance of redundant fits
- accessibility
- ease of assembly.

In designing for ease of machining it is important not to simply consider the functionality of the design but also to consider manufacturing process requirements. Although not exhaustive, Figs 5.8, 5.9, 5.10 and 5.11 give some examples which illustrate the principles involved in simplifying manufacture.

Figure 5.8 illustrates provision of runouts for cutting tools. A poor finish may result or a

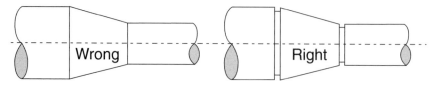

Figure 5.8 Provide for runout of cutting tool

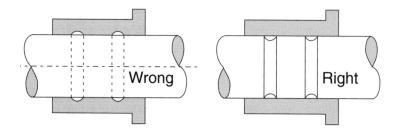

Figure 5.9 Machine features in easiest component

component may be impossible to manufacture without runouts. The decision to provide such undercuts must be taken at the design stage since the smaller diameter may have strength reduction implications.

The simplification of machining can also be achieved by the placing of features in the components which are easiest to machine and subsequently inspect. This is true of the grooves which are required for some sort of rubber sealing ring in the example shown in Fig. 5.9. If the grooves are designed to go on the inside of the housing then access for machining is difficult. If it is functionally acceptable to insert the grooves in the shaft then this should be done because turning the grooves in the shaft is much more straight forward.

In order to facilitate drilling, clearance must be provided for the drill to break through fully, drills should encounter equal resistance on their cutting edges so should not enter on a sloping surface and the centre of holes should be at least a full diameter away from the edge of a component to avoid breaking out.

The final decision between manufacturing methods is often based upon careful cost estimates of the alternatives. However, as a general guide choose simple shapes, such as cylindrical or flat surfaces. Try to avoid tapers and complex curves. Use the most economical machining process. Turning is much more economical than grinding. Keep machining to a minimum, as in the top diagrams of Fig. 5.10, by machining the feet only. The aim should be to only machine working surfaces and reduce the area to be machined. In the lower diagrams of Fig. 5.10 re-clamping or setting of tools during machining is avoided by designing machined features at the same height. Secure clamping is essential for accurate machining and care should be taken to ensure that faces, bosses and lugs are provided.

It is important to avoid redundant fits since they demand excessively tight tolerances. In Fig. 5.11, the bush which is inserted in the hole in the housing must have its depth of insertion controlled. This can be accomplished by the single 'flange' contacting the face of the housing, there is no need for a second register.

One of the goals during the embodiment stage should be to optimize the number of components. From the point of view of assembly, the smaller the number of components

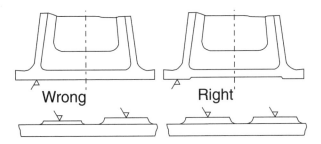

Figure 5.10 Minimize machining and re-setting

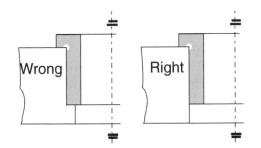

Figure 5.11 Avoid redundant fits

the easier assembly will be. However, reducing the number of components implies an increase in manufacturing complexity for those components. This is trade off, but the initial aim should be to ascertain the minimum number of components. If the answer to the following questions is no it is possible that components could be combined:

- Do the components move with respect to each other?
- Do the components have to be of different materials?
- Are separate components required for assembly and disassembly?

The aim must be to make assembly easier rather than just possible. Symmetry is not desirable and the introduction of features which break symmetry helps to ensure correct assembly. Assembly is greatly assisted by the addition of such features as lead-in chamfers.

5.7 Industrial design

It is very easy for engineering designers to believe that design is concerned only with achieving a technically optimum solution which will automatically satisfy aesthetic and ergonomic aspects. However, design is becoming ever more complex and if competitive advantage is to be gained the whole nature of the product must be addressed. The product development and introduction process is carried out by teams of 'experts' working cohesively towards a common goal. Such teams can only function well if each member is aware of the abilities and aims of his colleagues. The object here is to foster an understanding of the important role fulfilled by industrial design.

There are broadly three areas covering the complete range of design activity: technical, ergonomic and aesthetic. A very general distinction between engineering and industrial design is:

- The *engineering designer* is biased towards producing goods which have use.
- The *industrial designer* is biased towards ensuring useful products satisfy and appeal to their users.

The basic aims of industrial design are:

- Products must satisfy people in the ergonomic sense.
- Products should satisfy the natural human need for beauty, style and status.

Ergonomics

The subject of ergonomics is equally important to both types of designer and represents the overlap between industrial and engineering design. The word 'ergonomics' stems from two Greek words; 'ergos' meaning work and 'nomos' meaning the laws. In the USA 'human engineering' finds favour and in continental Europe the expression 'biotechnics' is often used.

Engineering ergonomics is concerned with ways of designing machines, operations and work environments to match human capacities and limitations. During the Second World War a new category of machine appeared which made demands on sensory, perceptual, judgemental and decision-making abilities rather than muscular power. This new class of machine posed some interesting questions about human abilities that could no longer be answered by common sense and time and motion studies. Consider the design of a manned space vehicle. Almost everything known about human beings is important: body dimensions; physiological reactions; sensory capacities; control abilities; eating, drinking and waste disposal; psychological effects of fatigue and emotion. Happily, for most designers, we are only usually concerned with a restricted part of the human sensory repertory.

The concern here is the design of man–machine systems, as illustrated in Fig. 2.6. Examples are the bicycle, motor car and manufacturing plant. All equipment systems are built for some human purpose and the vast majority of systems are monitored and controlled by humans. However, the degree of human involvement varies enormously. The human problems involved in the development of complex systems can be grouped into two different classes:

- those concerned with designing the system to complement the capacities and limitations of the human operators,
- those concerned with procurement, selection classification, training and promotion of people operating the system.

The distinction between these classifications of problem is not always clear and the designer must often make decisions as to the required level of skill of an operator. Design decisions must be made regarding the functions to be performed by the different parts of the man–machine system. The diagram is a simple representation of the role of a human being in a man–machine system. The human function is to sense an input, process the signal, perform some type of calculation, reach a decision and then perform some controlling action. The action is usually pressing a button or operating a lever which modifies the behaviour of the machine. These actions are performed within an environmental envelope, the effect of which on the operator cannot be ignored.

As an example of the 'possible' effect poor ergonomic design can have consider the inquest findings into the Kegworth air disaster. At 8.25 on the evening of 8 January 1989, a British Midland Boeing 737-400 crash-landed on the M1 close to Kegworth in Leicestershire. A fire had started in the left engine and the aircraft was subjected to severe vibration. The co-pilot, for whatever reason, considered the problem to be with the right engine and so shut that down. The left engine continued to function for a while but eventually failed, leaving the aircraft without power. The aircraft came down just short of the runway at East Midlands airport.

The vibration indicators are reproduced in Fig. 5.12 actual size. It is clear that they are small, particularly since they need to be read when the aircraft is vibrating. In all categories of work operator efficiency, speed and accuracy are influenced by the design of the components operated and the communication channels used to impart the information required for this operation.

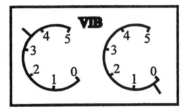

Figure 5.12 Vibration indicator

Clearly this is an oversimplification of the complex and stressful situation which confronted the pilots. However, the designer must ensure that optimum use is made of human operators and that optimum methods of communication of data are used. The following are some of the questions which should be addressed during the design of products and processes which involve human interaction:

- What role is the operator expected to play?
- Will optimum use be made of inherent human capacities?
- How will the equipment fit the operator?
- Will the operator sit or stand?
- Will the operator's posture be satisfactory?
- Is the section of population likely to operate the equipment clearly defined?
- What information does the operator need to perform the task?
- Should this information be visual, auditory or tactile?
- What type of display will give quick information with minimum ambiguity?
- What type of controls will be optimum?
- How much force can the operator be reasonably expected to apply?
- What form of communication is appropriate between operators?
- What physical and mental work will the operator be required to do?
- What are the ambient conditions likely to be?

Aesthetics

Confident recognition is dependent on the firmness of the visual statement. This is often undermined by the existence of an irrational factor in perception leading to overtones of doubt in identification. The avoidance of such doubt is the responsibility of the designer and is reason enough to justify the aim that all engineering designers have some knowledge of aesthetics.

Also, some knowledge of the workings of the eye is helpful, particularly of alpha rhythms. These are rhythmic oscillations at a frequency of about 10 per second which transform spatial patterns into temporal patterns. This transformation is roughly equivalent to scanning. Where the visual signals are indefinite, this scanning process is slow and confused. Thus clear visual form is our aim.

Confusion of visual form and the attributes attached can be caused by irrationality. In interpreting the well-known diagrams of Fig. 5.13 our eye and brain distort the images. In (a) the two vertical lines do not look the same length; in (b) the horizontal lines do not look straight; in (c) we often miss the repeated word; in (d) the horizontal lines do not look the same length.

This misinterpretation can operate in two main ways. The first occurs when the observed form strongly resembles other well-known forms. Certain qualities will be attached to the observed form, however improbable they may seem. The second is structural improbability, when physical defects may be assumed to exist in the form which is in fact structurally sound. Any engineering project evoking such a response, however irrational, will be less successful than the technical effort deserves. Aesthetic quality can have a real commercial significance in underwriting the total effort expended. Reference to the well-known illusory optical effects should remove any doubt that the brain does not always see the same as the eye.

When attention is either directed or attracted by any visual feature all other features tend to lose significance. This is called the 'figure on ground' effect and is fundamental to the perception and recognition of features. This effect is illustrated by the Runen vase shown

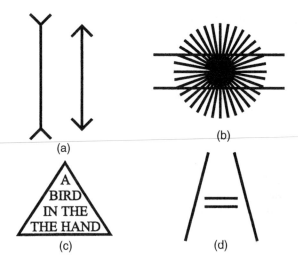

Figure 5.13 Optical illusions

Figure 5.14 Runen vase

in Fig. 5.14. At first sight most people see a white goblet against a black background. Eventually two human faces looking at each other are seen. Whatever is seen the other image or 'ground' tends to lose its identity.

The ability to perceive detail is called acuity and is mainly influenced by the level of illumination in relation to the amount of light reflected from the background. This is known as the brightness ratio and German Gestalt (form) psychologists have studied the influence on 'wholeness'. This is best illustrated using the diagrams in Fig. 5.15. In Fig. 5.15(a) the observer tends to automatically make sense of the random lines and sees the hexagon of Fig. 5.15(b). This whole seeking tendency may produce wholes where they do not exist or are not intended. Thus, since the mind is pattern seeking, the spots in Fig. 5.15(c) are also regarded as forming a hexagon.

The effect of colour is another aspect of aesthetics which is complex. It is apparent that all living organisms have a 'radiation' sense. In human beings this has been shown to be independent of conscious vision. Colour effects tend to be mainly in one of two directions, toward red or blue, and these two major colours induce different levels of activity in the brain. For example, conditions such as Parkinson's disease can be diminished by wearing green lenses, thus protecting from red. Brain activity is consistently greater under red light. Judgement is affected by colour. For example, time is overestimated in red light and underestimated in blue light. Light and colour density affect body functions just as they influence both mind and emotion.

Two simple conclusions are used by architects in designing the environment. With high levels of illumination and warm colours in the surroundings the body tends to direct its

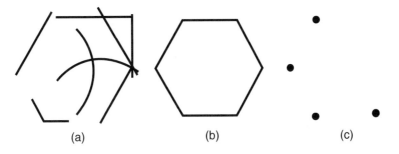

(a)   (b)   (c)

Figure 5.15 Wholeness illustration

attention outward. Such an environment is conducive to muscular effort and cheerful spirit. With softer surroundings, grey, blue and green, and lower brightness there is less distraction and a person can concentrate on mental tasks.

In the co-ordination of function and appearance the designer must search for interesting combinations of shape. Vitality of form often lies in the ability of certain properties to contain themselves or be contained time and time again. One such ratio, the *golden mean*, involves the division of a line into unequal parts such that the smaller is to the larger as the larger is to the whole. The ratio is repeated and it is this kind of repetition which creates and maintains interest. The golden mean or ratio is 1.618:1. Division at or about the golden mean often tends to produce greater aesthetic satisfaction than division elsewhere. It is the point at which a single figure can be placed most effectively on a long horizontal surface, as in Fig. 5.16(a), or the point at which the centre of bulk of two or more objects should fall, as in Fig. 5.16(b).

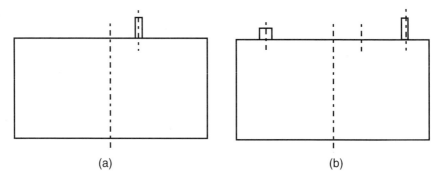

(a)   (b)

Figure 5.16 Golden mean

This can further be extended into the consideration of areas and is then referred to as the *golden rectangle* shown in Fig. 5.17. This rectangle has its length to height in the ratio 1:0.618. If the rectangle is divided by the vertical line ME, a square AMED and another golden rectangle MBCE are formed. If the diagonal AC is drawn to intersect ME at P, a further square MBQP and a third golden rectangle PQCE are produced.

It is not always possible or desirable to incorporate these particular proportions. Vitality, movement and interest can be obtained using alternative forms. Other forms capable of lending strength and beauty to appearance include the ellipse, hyperbola, parabola, the

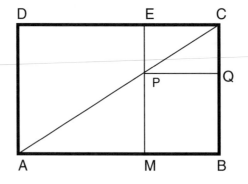

Figure 5.17 Golden rectangle

cycloids and involute. Simplicity, clean lines and good proportion are what the designer should aim for.

5.8 Principles

Embodiment principles

Optimization   The search is for the best compromise between conflicting criteria.

Simplification   Load transmission and form should be simplified.

Scaling   Full-scale models are rarely possible and reduced size testing must be carried out carefully.

Aesthetics   The designer should aim for a visually appealing product.

Ergonomics   A user friendly design is sought which makes appropriate use of the inherent skills of operators.

Synthesis   A solution is often arrived at by a combination of techniques and elements.

Iteration   Progress towards the detail design stage is made iteratively as knowledge of the important factors grows.

6 Modelling

There are three main methods of modelling covered in this chapter. Mathematical modelling, where equations are developed and tested within stated assumptions, is illustrated using the example of a pillar drill. Within this category of modelling it is sometimes possible to define equations which fully constrain the problem. The identification and subsequent solution of these equations leads to an 'optimum' solution. The second modelling method presented involves the creation of 2D and 3D scale models which are illustrated by means of a linkage mechanism and a wooden model of a car in a wind tunnel respectively. Finally, simulation using computers is discussed.

6.1 Introduction

Reference back to the engineering design process diagram, Fig. 1.5, reveals that there is no clear identification of a stage in the design process when modelling takes place. This is because it is a continual process which increases in complexity as the design proceeds. As mentioned earlier simplified models with many assumptions are made even before concept selection. Many more models are employed at the detail design stage. However, it is in the period between the selection of a concept and the commitment to continue into the relatively expensive, in terms of both time and money, detail stage that modelling is used widely.

There are three broad types of modelling. The mathematical model, including optimization, the physical scale model and simulation, which is normally carried out with the aid of a computer. In the widest sense a model may be defined as a simplified representation of reality, created for a specific purpose. Perhaps the most difficult part of the analysis of a new product or design is the formation of a realistic model.

A model, be it mathematical or otherwise, is only suitable for investigating a particular criterion or adequacy of behaviour. Consider for example the traditional bicycle wheel shown in Fig. 6.1. At first glance it may be tempting to model this as if compressive forces were acting through the lower spokes. However, compressive loads are avoided by careful design and a model based on compressive loading with a failure criterion of buckling would clearly be in error. In fact, the load F is transferred from the hub to the ground via the top spokes and the rim. The reason that such thin, and therefore light, spokes can be used is that they are designed to be in tension. Compression is actively avoided in the spokes by only fixing the spoke using a nut on the outside of the rim. The likely modes of failure are thus excessive tensile loading or fatigue (repeated cycles of loading which individually would not cause failure).

Ideally, every new or modified product being designed would be built and tested at full size. There are two main factors often preventing this, time and cost. In our increasingly competitive world the time taken to design, develop, test, manufacture and launch a new

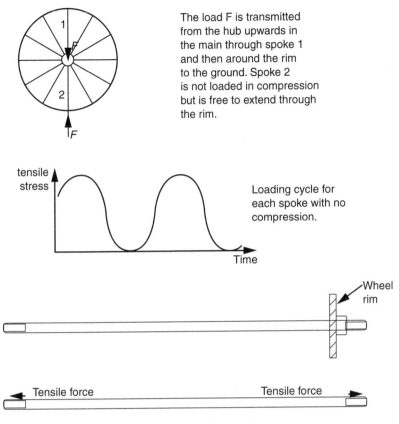

The load F is transmitted from the hub upwards in the main through spoke 1 and then around the rim to the ground. Spoke 2 is not loaded in compression but is free to extend through the rim.

Loading cycle for each spoke with no compression.

Free body diagram of the spoke

Figure 6.1 Simple analysis of wheel spoke strength

product has to be reduced to the minimum practical, particularly for consumer products. It is very difficult to capture market share from a competitor's previously launched product, almost irrespective of the improved design and features your product may include. Full size models would take a long time to manufacture and test delaying the subsequent launch and thus diminishing any competitive advantage of the new product. There are some exceptions, notably in electronic circuit design, where the behaviour of a complex circuit is often difficult to predict by calculation and the circuit is relatively cheap to build.

There are however many instances where the manufacture of full scale models during design is clearly impossible but where functioning must be verified during the design stages. Ships, aeroplanes, nuclear reactors and chemical plants are just a few examples of systems which involve the commitment of large capital expenditure and yet the designers must be convinced of at least a high probability of success. Such large systems are usually broken down into smaller sub-sections for modelling. In chemical plant, the process is first proved on a small scale in the laboratory. Another example would be the use of a small wooden model in a wind tunnel to predict properties such as drag and lift.

In many cases the model can be entirely on paper, in the form of drawings and mathematical representations, such as using differential calculus when modelling electrical circuits. Mathematical modelling is usually the cheapest and quickest modelling technique to employ, costing less than construction.

6.2 Mathematical modelling

Quotation – Ashley Perry's fifth statistical axiom.

> The product of an arithmetical computation is the answer to an equation. It is not the solution to the problem.

This quotation is presented as a caution. Mathematical and often computer based techniques can be used to solve many problems but the model of the problem on which any equations are based must represent the actual problem. Otherwise, the solution of the equations will mislead.

In the initial stages of a design, overall calculations are performed to verify working principles and to ascertain whether or not further work is justified. Should this not prove to be the case it is prudent that the concept should be rejected and either another selected or the project terminated before detail design work is considered, saving valuable time, effort and money. At the concept stage many assumptions are usually made and as long as these can be quantified the approach is justified. Most science based subjects are taught initially with many assumptions and simplifications in order to explain basic principles. More advanced theories are presented later. This is precisely the way a design problem should be tackled. Start simply with quick overall analyses, increasing the complexity later if the project warrants further investigation.

As the design process continues with more details being specified it is possible to reduce the number of assumptions being made and more accurate calculations can be performed. More realistic models are formed with fewer assumptions until a model which is as close as possible to the real situation is formed. The actual calculation procedure is normally relatively routine, the creative and intuitive process lies in the definition of the model. It is the definition of the model with its associated assumptions and potential modes of failure that is presented here.

There are some situations where the form of the model is dictated for one reason or another. For example, the specification of pressure vessels is governed by statute, the design of gears guided by standards and the selection of bearings governed by guidance in manufacturers' catalogues. Selection procedures for components from manufacturers' catalogues should always be followed since ignoring them would negate any guaranty. In such cases it only remains for the designer to manipulate the equations and interpret the results.

The fact that a simplified model is used probably means that it does not describe the reality precisely. However, the engineer is usually prepared to accept this limitation provided that the reality is described adequately for the purpose for which the model was constructed. As indicated, the first step in mathematical modelling is to replace the reality with a simplified model. This is usually accomplished by making simplifying assumptions to establish idealized components. Examples of such assumptions which should be familiar to all engineers are shown in Fig. 6.2.

Once the model is formed it is often necessary for the input data to be idealized, again by means of simplifying assumptions. As an example consider the loading encountered

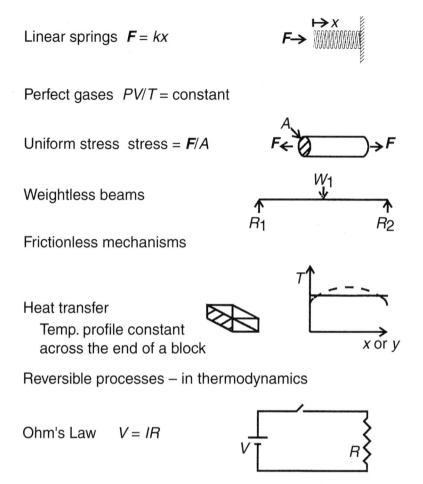

Linear springs $F = kx$

Perfect gases $PV/T = $ constant

Uniform stress stress $= F/A$

Weightless beams

Frictionless mechanisms

Heat transfer
  Temp. profile constant
  across the end of a block

Reversible processes – in thermodynamics

Ohm's Law $V = IR$

Figure 6.2 Common simplifying assumptions

during repeated or cyclic loading. The actual loading regime is often extremely complex but is simplified by ignoring the shape of the loading curve and considering only the number of times maximum and minimum values are reached. Fortunately, in empirical development of fatigue data this approach has been vindicated. Another data input often used in idealized form is that of material strength. Unique values are used which are impossible to achieve in practice and a spread of values exists for every material specification. In all cases the accuracy of any result is clearly limited by how well the simplified model describes the real situation and by the accuracy of the input data.

One of the major tasks confronting the designer is the definition of what constitutes failure and how to prevent it. Consider the example of the stress in a close coiled helical spring for a motor vehicle suspension. If the criterion used for failure is that the spring is only subjected to static loading then:

$$\text{Permissible shear stress} = 8PD/\pi d^3$$

where P is the load, D is the mean diameter of the coils and d is the wire diameter.

However, this mathematical model for permissible shear stress based on *static* loading would lead to an under estimate of the required material strength. It is very clear in this case that the criterion used for failure should be based on *dynamic* loading and that the mathematical model must indicate the ability of the spring to withstand the worst type of loading. The responsibility of the designer is to identify all the possible modes of failure of any machine or structure and to establish the models accordingly. The following list gives most possible mechanical failure modes, starting with the more common.

* elastic deformation
* yielding
* creep
* brinelling
* fretting
* scoring
* thermal relaxation

* fatigue
* corrosion
* erosion
* ductile fracture
* galling
* stress rupture
* oxidation

* wear
* impact
* buckling
* brittle fracture
* seizure
* thermal shock
* radiation.

A *four stage* approach to forming a *mathematical model* is recommended:

(1)   Draw three separate diagrams, one showing the intended geometry, a second indicating the applied forces and moments and a third indicating the resultant forces and moments on the critical section(s).

(2)   State all assumptions.

(3)   Develop mathematical equations for the resultant forces and moments at the critical section(s) in terms of the applied loading and the geometry.

(4)   State the failure criteria.

Example

Consider the mathematical modelling of the column for a pillar drill as shown in Fig. 6.3. Each part of the drill must be examined for strength. Presented here is the modelling of just one critical section as indicated X–X in Fig. 6.3(a).

Diagrams
The applied forces and moments are identified in Fig. 6.3(b) as the force applied on the handle by the operator F, a reaction F at the drill tip which is in direct proportion to the force applied by the operator, Mg due to the drill motor, mg due to the drill arm and a moment M due to the cutting by the drill. Figure 6.3(c) shows the resultant forces and moments at the critical section. These are the combination of vertical and horizontal forces together with twisting and bending moments.

Assumptions
(1) The drill is rotating at constant speed.
(2) The rate of feed into the workpiece is constant.
(3) There is no deflection in the drill arm.
(4) Force applied horizontally represents the worst case.

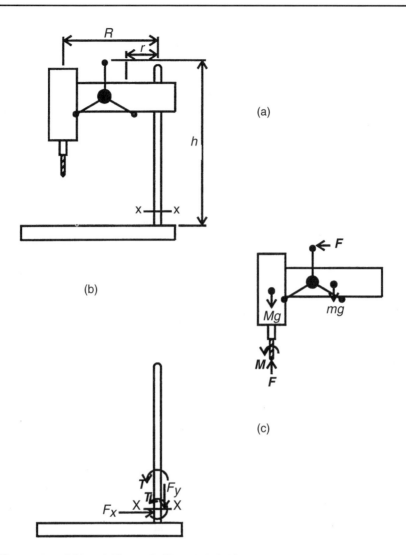

Figure 6.3 Pillar drill modelling. **(a)** Pillar drill layout and geometry. **(b)** Applied forces and moments. **(c)** Resultant forces and moments acting on section X-X

(5) There is no friction in the bearings.
(6) The pillar above the critical section is weightless.
(7) There is no sideways loading.

Resulting forces and moments

Vertical force	$F_v = F - Mg - mg$
Horizontal force	$Fh = F$
Bending moment	$t = Fh + MgR + mgr - FR$
Twisting moment	$T = M$

Failure criteria

(1) The maximum principle stresses occurring at the cross-section must not exceed the yield stress of the material. These stresses will be caused by a combination of compression (or tension), bending and torsion.
(2) The column must not buckle.
(3) Deflection should always be within acceptable limits. Any inaccuracies caused by the column bending will affect drill accuracy.
(4) The natural frequency of vibration should be well away from the operating frequency of the drill.
(5) Since the loading is cyclic, fatigue life should be estimated.

The above procedures should be repeated until the worst case loading is discovered. For example, it is obvious that the compressive loading on the column would increase if the load *F* were applied vertically. Also the identification of critical sections to model must be made with care. In most structures the critical areas are joints, not the major components. This is often because the 'flow' of the stresses is disturbed and stress levels are raised locally. In a simple pin joint, for example, the components being joined are drilled to accept the pin. This inevitably weakens the components due to the removal of material.

Once the model is established the full power of mathematics and computer analysis can be brought to bear. However, the fact that the most brilliant manipulation of a poor model will lead to inaccurate and potentially misleading results cannot be over emphasized.

6.3 Optimization

A common characteristic of design problems is the existence of conflicting requirements. In general we cannot have long life and high efficiency and greatest number of features and fewest moving parts, for example. However, some requirements do go together such as small size generally coincides with lightest weight and fewest parts often means greatest reliability. The selection or design of the best possible solution depends on a clear definition of the interaction of all the pertinent variables affecting the problem, an explicit statement of the design objective and an effective procedure for locating the optimum solution in accordance with the stated objective.

Optimization is the process of determining the values of the variables, subject to various constraints, that make a desired criterion a maximum or minimum. A common criterion is cost or weight. However, the criterion can be any property or ratio of properties relevant to the design, such as power to weight ratio. An optimization problem usually involves three types of functional relationships among the specifications and design parameters. These are:

Criterion function This is the mathematical expression of the quantity whose maximum or minimum is to be found as a function of the design parameters. This criterion may be single or a ratio of several characteristics. There can only be one criterion function.

Functional constraints These equations are the physical laws involved in the proposed design. They constitute the mathematical model as previously described. The number of equations in this set must be less than the number of design parameters.

Regional constraints   There is no limit to the number of regional constraints and they are always expressed as inequalities. These functions are mathematical statements of the limits between which design parameters must lie.

There are many methods of optimization including linear programming, differential calculus, dual variables and geometric programming. It is beyond the scope of this text to include details on each of these but example problems are presented which have been solved by the first two methods.

Optimization by linear programming

This is a method which can be applied in the solution of problems in which the criterion function and the constraints are linear functions of the variables. When there are only two or three variables the problem can be solved graphically. The extremes of satisfaction of the equations are plotted and as a result a feasible region is identified. The optimum point lies along the boundary of the feasible region and which point is dictated by the slope of the line representing the criterion function.

As an example consider the suggested design of a uniform column of tubular section as shown in Fig. 6.4(a). The column is to carry a compressive load (P) of 25 kN at the

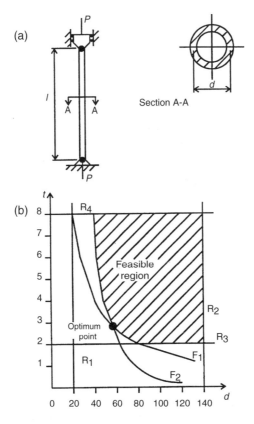

Figure 6.4 Optimum column design. **(a)** Column geometry and cross-section. **(b)** Graphical solution

minimum overall cost possible. The column material has already been selected and has a Young's modulus of 85 GN/m², a yield strength of 50 MN/m² and a density of 2500 kg/m³. The length of the column is 2.5 m and the mean diameter of the tube is restricted between 20 and 140 mm. Wall thicknesses outside the range 2 to 8 mm are not available. The cost of the column, including material and construction costs can be taken as $5M + 200d$ where M is the mass and d is the mean diameter of the column.

The criterion function to be minimized is $C = 5M + 200d$. It is necessary to express this and all subsequent equations in terms of the variables of the column. These variables are diameter d and thickness t. Hence the criterion function becomes

$$CF = 5M + 200d = 5\rho\pi dtl + 200d$$

There are two functional constraints since the column could fail due to either compressive (yield) stress or it could buckle. Considering yield,

$$\text{Induced stress} = P/\pi dt$$

Since the limiting yield strength is 50 MN/m² the first functional constraint is

$$F_1, P/\pi dt - 50 \times 10^6 \leq 0 \qquad dt \geq 159 \times 10^{-6}$$

Considering buckling,

$$\text{Euler crippling load} = \pi^2 EI/l^2$$

where

E is Young's modulus
l is the length of the column
I is second moment of area, $I = \pi dt(d^2 + t^2)/8$

Since the buckling stress is required and not the load then we divide by the area, giving bucking stress $= \pi^2 EI/l^2 \pi dt$. Young's modulus for the material is 85 GN/m² so the second functional constraint is

$$F_2, P/\pi dt - \pi^2 E(d^2 + t^2)/8l^2 \leq 0 \qquad dt(d^2 + t^2) \geq 473.7 \times 10^{-9}$$

There are also two regional constraints which relate to the diameter and wall thickness restrictions. They are

R_1 $20 \leq d \leq 140$

R_2 $2 \leq t \leq 8$

Since there are only two design variables, d and t, the problem can be solved graphically as shown in Fig. 6.4(b). The axes are d and t and first the regional constraints are plotted, leaving a rectangular feasible region between $d = 20$ and 140 mm and $t = 2$ and 8 mm. Next the functional constraint curves, F_1 and F_2, are plotted. The cross-hatched area indicates the feasible region within which all potential solutions must lie.

In order to determine the optimum design point within this region the criterion function is plotted. The slope of the line is what matters, and since the aim is to minimize cost this line is moved over the feasible region towards the origin until the last point of contact with the feasible region is determined. This is seen to occur at $d = 54\,mm$ and $t = 3\,mm$. However, a 54 mm diameter tube is not readily available so the next size up is used, $d = 55\,mm$ and $t = 3\,mm$.

In the graphical solution of linear equations two variables can be handled easily. Three variables can be plotted on 3D graph paper. Four or more variables takes us beyond the realms of graphical solution and an algebraic solution is necessary. The Simplex method has been developed for this purpose and when computerized is straightforward to use. It is a repetitive procedure which involves moving systematically, one step at a time, to the next intersection on a feasibility polygon until an optimum is discovered.

Optimization by differential calculus

Often there are no functional constraints and it is possible temporarily to ignore any regional constraints. In such cases, assuming the criterion function is a differential expression, differential calculus can be used to determine the optimum. With this method the optimum is determined by the solution of the simultaneous equations found by setting the derivatives of the criterion function with respect to each of the parameters to zero. In the case of one parameter this is equivalent to finding the point where the slope is zero. Thus, maximum or minimum (optimum) values are calculated.

Consider the example of an electrical power cable. Electrical power P is to be supplied at a direct current voltage V over a distance S, with a return cable being used. The conductors are of specific resistivity p and cost c_1 per unit volume year. Power costs are c_2 per unit year. Find the equation governing the minimum cost of transmitting the power.

$$\text{Total cost} = \text{material cost} + \text{power loss costs}$$

$$\text{Material costs} = c_1 \times 2SA$$

where A = cross-sectional area.

$$\text{Power loss costs} = c_2 \times I^2 R$$

Since $I = P/V$ and $R = pS/A$

$$\text{Power loss costs} = c_2 \times (P^2/V^2) \times (2Sp/A)$$

Criterion function,

$$C = c_1 \times 2SA + c_2 \times (P^2/V^2) \times (2Sp/A)$$

Differentiating,

$$\frac{dC}{dA} = c_1 2S - c_2 \times (P^2/V^2) \times (2Sp/A^{-2}) = 0$$

Hence,

$$A = P/V'pc_2/c_1$$

Multiplying each term in the differentiated equation by A gives the minimum cost as

$$2Sc_1 \frac{P'}{V} \frac{pc_2}{c_1} + 2Sc_2 \frac{P^2}{V^2} \frac{pV'c_1}{Ppc_2}$$

Which can be simplified to

$$\text{Minimum cost} = \frac{4SP'pc_2c_1}{V}$$

As already mentioned this section covering optimization is not intended as a full treatise nor is there sufficient information to enable an engineer to become expert. All that is intended here is an illustration of the possibilities for the modelling of a problem mathematically which leads to a guaranteed optimum solution within stated assumptions. As a final example consider the following heat exchanger pipe packing problem.

In the design of fluid flow heat exchangers the length of pipe in the exchanger influences the rate of heat exchange. The cost of the pipe is a governing factor in the optimization process, along with the cost of the tank shell within which the pipes are located and the temperature altering fluid flow. The shell can be fabricated in a variety of shapes: a cylinder which houses most pipes for a given perimeter; a rectangle which may occupy less floor space than a cylinder; or some other shape which is easy to fabricate.

Consider the cost of a cylindrical heat exchanger tank of diameter D and length L that requires a minimum of 91.5 m of pipe to meet heat exchange requirements. The requirement of 91.5 m is itself an optimization process dependent upon heat flow rates, material heat capacities, incoming fluid temperature, required outgoing temperature and cost and availability of pipe material.

Assume the cost of the tank shell has been estimated as the sum of the pipe costs at £700, the shell cost as $1590D^{2.5}L$ and the floor space cost as $215DL$:

$$C = 700 + 1590D^{2.5}L + 215DL \tag{1}$$

Additional data is that 20 pipes will fit into a 0.093 m^2 cross-section tank, giving the following analytic restriction of the problem $[\pi D^2/4][(L)(20 \text{ pipes}/0.093 \text{ m}^2)] > 91.5$ m which can be simplified to:

$$\pi D^2 L > 1.7 \text{ m} \tag{2}$$

From Equation (2), $L > 1.7/\pi D^2$. Substituting this in (1) gives

$$C = 700 + 1590D^{2.5}1.7/\pi D^2 + 215D1.7/\pi D^2$$

$$C = 700 + 860.4D^{0.5} + 116.4D^{-1}$$

$$dC/dD = 430.2D^{-0.5} - 116.4D^{-2} = 0$$

Hence, $430.2D^{1.5} = 116.4$

Therefore $\underline{D = 0.418 \text{ m}}$ and substituting in (2) gives $\underline{L = 3.1 \text{ m/pipe-length}}$

Substituting in (1) gives the optimum $\underline{\text{cost} = £1535.4}$

It is important to analyse whether this is a realistic result. If the length of each pipe is 3.1 m and 91.5 m are required then 29.51 pipes are required. That is impossible and 29 or 30 should be used. Twenty-nine pipes requires $L = 91.5/29 = 3.155$ m and $D = 0.414$ m giving a cost of £1534.0. Thirty pipes gives a cost of £1534.4. The three costs are approximately the same and reflect a system which is insensitive to the variables. The variation in cost (£1535.4 to £1534.0) indicated is likely to be within the accuracies of calculation and manufacturing tolerances of the complete system.

The commercial availability of the length of pipe could be another factor affecting the design. If material comes in 20 ft (6.097 m) lengths, then $L = 3.05$ is a good choice and 30 pipes are required. If pipe is supplied in 24 ft (7.317 m) lengths then additional analysis would be required to examine the effect of $L = 3.659$ m. The widest material the roll machine can handle, say 2.5 m, may also limit the design.

A larger problem is associated with the assumption that the ratio of the number of pipes per square unit is constant regardless of the area. Consider enclosing a number of pipes of diameter two units, in a cylindrical shell as shown in Fig. 6.5. If one pipe is enclosed in a cylindrical shell the area of the shell is π unit^2 and the pipe density is i pipe/π unit^2 = 0.32 pipes/unit^2. This is the largest possible pipe density because there is no wasted space. If two pipes are enclosed in a cylindrical shell the shell cross-sectional area is 4π unit^2 and the pipe density is 2 pipes/4π unit^2 = 0.16 pipes/unit^2, half as efficient as using one pipe. With three pipes the pipe density is 0.21. Continuing analysis shows that 1, 7, 13, 19, 31, 37, ... are the most efficient numbers of pipes.

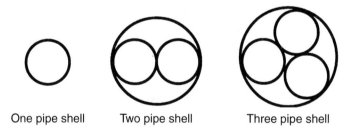

One pipe shell   Two pipe shell   Three pipe shell

Figure 6.5 Pipes in cylindrical shell

This being the case then the number of pipes in the shell in this example should be 31 and $L = 91.5/31 = 2.95$ m. Perhaps the optimum solution is to use thirty-one 3.05 m pipes and increase the safety margin by using 94.51 m of pipe rather than 91.5 m.

Optimization problems do not always yield a single solution!

6.4 Scale models

Often the most difficult stage in the engineering design process involves the verification of the likely behaviour of the design before the commitment is made to detail the design.

Mathematical models are an important first step but the power of the development and use of two- and three-dimensional scale models cannot be over stated.

Two-dimensional models include mannequins for assessing ergonomic suitability, models for assessing the behaviour of linkage mechanisms, photoelastic models for visualization and quantification of stresses and strains, and many others.

Mannequins are available in many different scales and many different sizes representing particular sections of the population. However, they should only be used as a general guide for assessing overall functionality of the man–machine system and should be supplemented by comprehensive fitting trials in full-scale models of the equipment. These three-dimensional models are still considerably cheaper to build than actual machines and are easier to modify, enabling more layouts to be tried. Modern flight simulators are examples of highly sophisticated full-size models which, as well as assessing ergonomics, integrate aesthetics and technology whilst being used for their prime objective of training pilots relatively cheaply.

A linkage mechanism model is often used to establish two main factors, the functionality of the mechanism and the space requirement of the moving links. Kits are available which consist of a mounting board onto which are fixed rigid links with holes along their axes through which pins can be inserted. One of the best known four bar linkages, devised by the mathematician Chebyshev, is shown in Fig. 6.6. The link ratios are as originally proposed and the coupler curve traced includes an approximate straight line with a quick return.

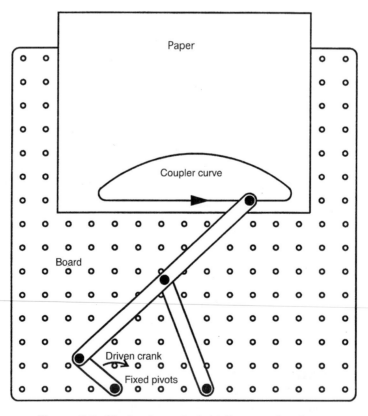

Figure 6.6 Chebyshev straight line mechanism

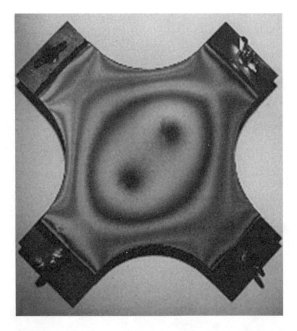

Figure 6.7 Star shape undergoing photoelasticity analysis

Even simple mechanisms such as that illustrated are more easily visualized using a model than by simply drawing the extremes of motion in a scheme drawing. Another advantage of the model is that the effects of altering the geometry can quickly be assessed. The mechanism shown has many applications such as in reciprocating cutting on a shaping machine and constant torque lifting with the straight line portion vertical.

Figure 6.7 shows a two-dimensional star shape undergoing photoelastic analysis. Photoelasticity is a visual, full-field technique. When a photoelastic material is strained and viewed with a polariscope, distinctive coloured fringe patterns are seen. Interpretation of the pattern reveals the overall strain distribution. The fringes which can be seen indicate areas of constant and relatively high stress. From such an assessment decisions can be made, such as strengthening areas where stress concentrations occur. It is also possible to quantify both stress and strain levels by counting the fringes as the load is increased. The photograph in Fig. 6.8 illustrates the fringes in a two-dimensional combustion engine connecting rod. Whilst this is clearly very idealized, ignoring as it does significant dynamic and inertia forces, regions of relatively high stress are clearly indicated.

Three-dimensional scale models may be used as a convenient way of obtaining a wide range of information which could otherwise only be gained by means of a great number of sketches and drawings or the manufacture of full-scale prototypes. They can be used to establish various kinds of experimental data, such as deflections, strength, drag coefficients and vibration characteristics. Such models, even with the advent of very sophisticated computer-based visualizations, remain the best medium for communicating ideas, within the engineering departments, to the rest of a company and particularly to outside clients.

The main benefit of constructing scale models as opposed to manufacturing prototypes is one of cost. The factors which can be established using scale models can be divided into three broad categories, technical, ergonomic and visual.

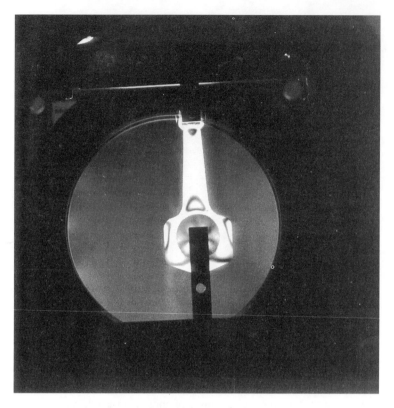

Figure 6.8 2D connecting rod

6.5 Simulation

Simulation techniques are mainly computer-based and some allow dynamic analysis. One of the more widely used techniques is finite element analysis. A finite element analysis of the head of a garden hoe is presented in Fig. 6.9 which, in a similar way to the photo-elasticity analysis, clearly indicates the highly stressed regions. Again, as with all modelling, the output from such a computer-based system is only as representative of the true situation as the model on which it is based. To a certain extent it is more critical that models are carefully formed when using a finite element package since the workings behind the solution cannot be readily understood by the user.

The basic idea in finite element analysis is to replace the complicated problem by a simpler one. The solution region is considered to be built up of many small interconnected regions called finite elements. In each element an approximate solution is assumed and the conditions of equilibrium for the structure are derived. The method can be applied to structural, heat transfer and fluid flow problems.

A further category of modelling which has been made possible with the rapid increase in speed of response and memory capacity of computers is synthesis. Synthesis is perhaps best explained by comparison with design analysis. When employing analysis techniques

Figure 6.9 Finite element analysis of garden hoe

the designer is trying to calculate how a preselected design will function. These calculations could involve establishing stress levels, deflections, reliability, or many other parameters, but very few, if any, design changes will be suggested. Synthesis on the other hand attempts to develop something which did not exist before by the use of different materials and sizes of configurations. In analysis factors are used as constants, in synthesis they are treated as variables.

6.6 Principles

Modelling principles

Optimization   The search is for the best compromise between the conflicting criteria.

Simplification   In order to model a situation it is first necessary to make some simplifying assumptions.

Scaling   Full-scale models are rarely possible and reduced size testing must be carried out carefully.

Visualization   Computer generated or physical models which aid visualization of the final product are very useful.

Synthesis   A solution is often arrived at by a combination of techniques and elements.

Iteration   At all stages the models will need to be used iteratively as knowledge of the important factors grows.

6.7 Exercises

1. The schematic drawings show devices which require to be investigated using mathematical models in order to determine the adequacy of each component or section to perform its intended function. The model for each of spanner, bicycle crank, golf club and grinding wheel must include:

- three separate diagrams, one showing the intended geometry, a second indicating the applied forces and moments and a third indicating the resultant forces and moments on the critical section(s)
- all assumptions
- mathematical equations for the resultant forces and moments at the critical section(s) in terms of the applied loading and the geometry
- the failure criteria.

At this stage of the design process, you are not being asked to solve for each of the sections. All that is required is the forming of the model, which is the critical starting point for any analysis of the intended design.

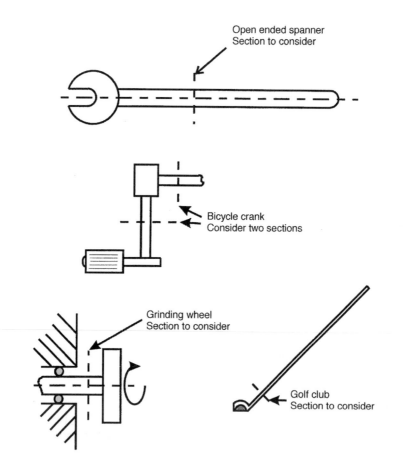

Open ended spanner
Section to consider

Bicycle crank
Consider two sections

Grinding wheel
Section to consider

Golf club
Section to consider

2. A pole of height H on which a horizontal pull of F (N) is applied is to have a wire-rope guy attached in the plane of pull, as shown in the figure. The minimum allowable diameter of the wire-rope depends upon its strength σ_u, the angle θ and the load F. If the cost of the wire-rope is a function of its volume, determine the angle for minimum wire-rope cost.

If the cross-sectional area of the wire-rope is $0.38D^2$ (not $\pi D^2/4$) determine the diameter of the wire-rope as a function of F, θ and σ_u. The steel wire-rope is supplied in diameter increments of 3 mm in the range 6 mm $< D <$ 24 mm. The wire-rope strength, $\sigma_u =$ 520 MPa and a factor of safety of 3.5 is appropriate. Assuming that the pole is 6 m high and that the force F is 4.5 kN determine the angle θ, the wire diameter and the length to use for minimum cost.

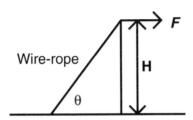

3. A helical compression spring is to be designed for minimum weight. The spring is to have a mean diameter of coils (D) of 30 mm and must deflect (δ) 13 mm under the maximum load (P) of 500 N. The torsional modulus of elasticity (G) for the material is 83 000 MN/m² and the number of inactive coils (Q) is 1.5. It can be shown that the optimization problem reduces to minimizing

$$W = (\pi^2 \rho \delta G d^6 / 32 P D^2) + (\pi^2 \rho Q D d^2 / 4)$$

subject to the stress constraint $16 P D^{0.75} / \pi \tau d^{2.75} < 1$

Using differential calculus calculate the wire diameter (d) which will give minimum weight.

4. Shown in the figure is a suspension bridge of span s and central sag h. The deck has constant weight per unit length w and the side spans are of length $s/2$. Since the cables hang in a parabolic shape the tangents to the cables at the tower tops, PR and PS, intersect a distance h below the centre of the cable. If the towers are made taller they become dearer

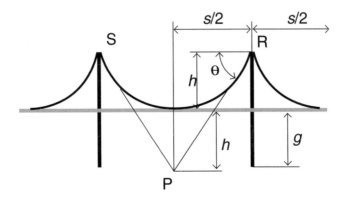

and the cables become cheaper since it can be shown that the tension in the cables and so their cross-sectional area is inversely proportional to h. Assuming the sag in the cable to be small compared with the length and height of the bridge write down an equation for the total cost of cables and towers, ignoring manufacturing and labour costs.

Suppose $\sigma_c = 600$ MN/m^2, $\sigma_t = 120$ MN/m^2, $c_c = £1200/m^3$ and $c_t = £2400/m^3$ are the allowable stresses and cost per unit volume of the cable and tower materials. Neglecting the weight of the cables and hangers calculate the optimum h/s ratio.

7 Detail design

Statistical assessment of the probability of failure is presented as an alternative to the use of factor of safety. The emphasis is on a better understanding of the limiting factors associated with the design. Quality and reliability are presented together and distinction drawn between the two. The most common mode of failure for structural components is by cyclic loading or fatigue. The chapter includes a section covering avoidance of stress concentrations using pipe flow analogy. High cycle fatigue life prediction is the main subject matter.

7.1 Introduction

Following the embodiment stage the next stage is to consider individual components and ensure that the design or selection of these is optimized.

During the detail design process the design and selection of each component is verified and information prepared which will enable manufacture to commence. The input to the detail design stage is the scheme drawing and the design intent. As in all other stages, all decisions must be made within the constraints of the PDS. The output is a series of production drawings accompanied by documentation. Again solutions must be synthesized and decisions made in the design of one component will influence the design of others.

The detail design process is illustrated in Fig. 7.1. As with all other stages it is cyclical or iterative in nature following broadly the pattern indicated in the outer ring of the figure.

7.2 Factor of safety

All components carry a load of some kind, be this load electronic, chemical or structural. As a critical part of our design calculations a designer must ensure that all components can sustain the applied load for the working life of the product or process. In simple terms structural failure can occur due to breakage or significant deformation and the analysis process can be seen in terms of the three stage process outlined in Chapter 6 'Modelling'.

(1) load type and force analysis;
(2) stress analysis of critical sections;
(3) analysis of possible modes of component failure.

The vast majority of failures are caused by dynamic or cyclic loading leading to fatigue failures. We are only concerned with static loading here, fatigue being covered in the later robust design section of this chapter.

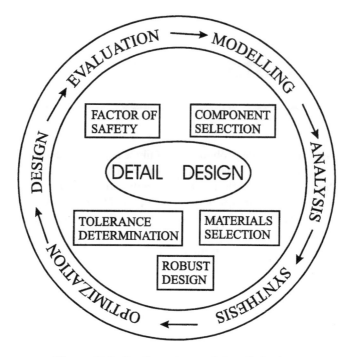

Figure 7.1 Cyclic nature of detail design

In the early stages of detail design the shape and form of a component are often ill defined so, with incomplete knowledge, we introduce a factor of safety to enable the analysis process to commence. Factor of safety is usually based upon the yield strength of the material and in exceptional circumstances upon the ultimate strength of the material.

The general equation for factor of safety is

$$\text{Factor of safety } (N) = \frac{\text{Load carrying capability}}{\text{Applied load}}$$

$$N = \sigma_Y/\sigma_L \quad \text{or} \quad N = \sigma_U/\sigma_L$$

Where, σ_L is the stress due to the applied load;

σ_Y is the yield strength of the material;

σ_U is the ultimate strength of the material.

In the first iteration we consider the nominal stress at the critical section to be the maximum applied stress. Subsequently, it is common practice to use the local maximum stress caused by such stress raisers as notches, shoulders, threads, holes, radii and undercuts. Such effects can be quantified using the many charts commonly available.

For ductile materials (< 1500 MN/m^2 tensile strength) typical factors of safety N are shown in Table 7.1.

Table 7.1

	Steady loads	*Occasional shock loads*
Tension and/or bending	3	6
Compression and/or contact pressure	3	6*
Torsion and /or shear	4.5	9

*This represents the crushing limit. You should also check buckling.

Since factors of safety compensate for uncertainties it is the designer's duty to gain further knowledge through investigation and research in order that doubts may be removed and lower factors of safety used.

Statistical assessment of factor of safety

The estimation of how reliable a product will be can be equated with a doctor taking a person's pulse rate. A high or low pulse rate indicates the patient is not well but does nothing to make the patient better. Similarly, estimation techniques will not improve the reliability of a product, merely indicate potential problem areas. Most texts concentrate on the very important area of tests for reliability estimation which are used in quality assurance and for scheduling preventative maintenance. However, the real issue confronting the designer is building in product reliability, in which case approximate reliability estimates will suffice and greater emphasis on reliability improvement techniques is required.

The situation facing design engineers is that most of the quantities upon which calculation procedures are based assume single figure values. However, it is not possible for materials, for example, to be produced economically without variation of properties. Variations also occur due to manufacturing methods such as heat treatment, weld quality, surface finish and dimensional accuracy. All other factors, such as fatigue life, fracture toughness, notch sensitivity, creep, abrasion and corrosion exhibit scatter.

The scatter exhibited does not always follow a normal distribution curve. However, for simplicity, in the later examples all data is assumed to follow the normal distribution. There are many different statistical distributions, some others are illustrated in the reliability section, which may in certain circumstances be more applicable. However, the principle here is that all values exhibit scatter and that quoting and using single figures is misleading. The reader must decide which distribution is most applicable to a particular situation. If a statistical distribution can be identified which fits the situation then a figure can be obtained, giving a more accurate representation of the situation.

For both applied load and load carrying capacity distributions it is normal to consider the area of the normal curve contained by six deviations. Indeed in engineering in general, including quality control, a tolerance specified is assumed to be three times the standard deviation. Taking three deviations a side covers 99.73% of all possibilities. The designer sets the allowable, acceptable chance of failure in the range of well-defined probabilities.

The concept of a factor of safety has been employed in engineering design for many years and although the necessity of its usage is well recognized the basis of its selection is often nebulous. Indeed, in some industries it is referred to as the factor of ignorance!

In calculating the factor of safety as $N = \sigma_Y/\sigma_L$ no account is taken of the shape of either the load distribution or the material strength distribution. Only mean values are used.

Clearly, the aim is to prevent overlap of the two distributions, therefore preventing failure. However, for a machine or structure it is impossible to predict precisely the external loads to which it will be subjected. Hence, in predicting the external load on an element, a tolerance band will accompany the specification of the mean load. Similarly, the load carrying capability of the element is affected by material strength variations and geometrical tolerances, both of which are susceptible to uncontrollable manufacturing flaws which can be significant. Thus the load carrying capability will be accompanied by a tolerance band.

The specific relationship which should exist between the two distribution curves for a satisfactory design will depend on the particular element under consideration and the significance attached to the occurrence of a particular failure mode. There are many options open to the designer once the likely distributions are known. To minimize the chance of failures occurring there are two main options, tighter manufacturing control and material selection reducing the spread of the capacity curve and increasing the separation of the mean values. Both of these will add cost to the design and an optimum or compromise solution must be arrived at.

Where failure would endanger human life, for example, the relationship shown in the figure would be desirable, since there is no realistic overlapping of the curves. However, there are many design situations where the occurrence of a particular failure mode could be tolerated occasionally with no significant consequence. For such cases the relationship between the distribution curves can be depicted as in Fig. 7.2.

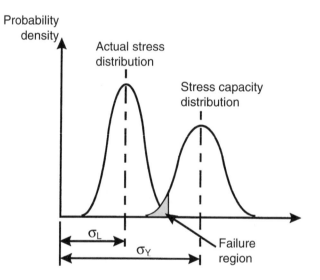

Figure 7.2 Load and strength distributions

For a component the particular failure mode will occur if the load capability based on that phenomenon is less than the corresponding actual load. Hence the distribution curve for the difference between load capability and actual load will be very significant since the probability of failure can be estimated from it. Depicted in Fig. 7.3 is the distribution curve for $(\sigma_Y - \sigma_L)$ where all negative values correspond to the occurrence of failure.

For cases where failure cannot be tolerated, the distribution curve $(\sigma_Y - \sigma_L)$ should be located so that all realistic values are +ve. When failure can occasionally be tolerated, some values can be −ve and the area under the curve represents the predicted percentage of

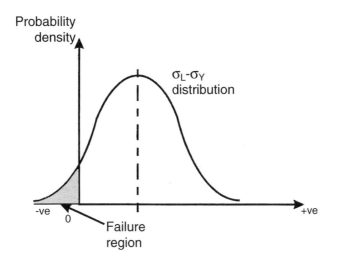

Figure 7.3 Combined curve

failures. For both loading and load capacity curves it is normal to consider six standard deviations, covering 99.73% of all cases. Practical distributions tend to exhibit a much more extended 'tail' than represented by the normal distribution. However, quality control of material or overload protection of a device are examples of methods employed to abruptly cut off the tail, making the use of the normal distribution valid.

In order to take advantage of available statistics tables a new unitless variable t is introduced.

$$t = \text{safety margin}$$

Any value of t can be converted to the proportion of the area under the normal curve between the mean ordinate and the ordinate at any standardized deviate from the mean using the normal table, reproduced in part in Table 7.2. The area shaded in the diagram is the area indicated from the tabular values. In order to obtain the full reliability figure we must add the more positive half of the area under the curve to the shaded area. Since we are only interested in high reliability figures only the higher values of t are quoted.

Since we are only interested in the point on the combined applied and capacity curve beyond which values go negative,

$$t = (\sigma_{\text{YMEAN}} - \sigma_{\text{LMEAN}})/((D_Y)^2 + (D_L)^2)^{1/2}$$

Consider the example of a component which has a load carrying capacity which is normally distributed, with a mean value of 5000 N and a standard deviation of 400 N. The load it has to withstand is also normally distributed, with a mean value of 3500 N and a standard deviation of 400 N.

The reliability of the component per load application is estimated by calculating the safety margin.

$$t = (5000 - 3500)/((400)^2 + (400)^2)^{1/2}$$

$$= 2.65$$

Table 7.2

t	0.00	0.01	0.02	0.03	0.04	0.05	0.06	0.07	0.08	0.09
2.0										
2.1			4840							
2.2										
2.3										
2.4										
2.5	4938	4940	4941	4943	4945	4946	4948	4949	4951	4952
2.6	4953	4955	4956	4957	4959	4960	4961	4962	4963	4964
2.7	4965	4966	4967	4968	4969	4970	4971	4972	4973	4974
2.8	4974	4975	4976	4977	4977	4978	4979	4979	4980	4981
2.9	4981	4982	4982	4983	4984	4984	4985	4985	4986	4986
3.0	4987	4987	4987	4988	4988	4989	4989	4989	4990	4990
3.1	4990	4991	4991	4991	4992	4992	4992	4992	4993	4993
3.2	4993									
3.3	4995									
3.4	49966									
3.5	49977									
3.6	49984									
3.7	49989									
3.8	49993									
3.9	49995									

Probability density

σ_L-σ_Y distribution

-ve   0   mean   +ve

Referring to the section of normal table a value for t of 0.496 is obtained. This must be added to the remainder of the positive area under the curve, since 0.496 only represents the shaded area of the diagram. Thus we have a reliability figure of 0.996 which would normally be expressed as 99.6%.

Extending the example, it is illuminating to examine the effect of larger standard deviations (less control of parameters). Consider standard deviations of 500 N.

$$t = (5000 - 3500)/((500)^2 + (500)^2)^{1/2}$$
$$= 2.12$$

Reliability drops to 98.3%.

Note that reliance on the static factor of safety alone would be misleading since it is calculated on mean values and would be unchanged.

7.3 Selection procedure for bought out components

Many components and units which form part of a product or system can be bought from companies specializing in their manufacture. Generally in the design of any system the successful selection of suitable elements is the result of matching system requirements with the capabilities of one of a wide range of available alternatives. Thus information about the system and information about available hardware is necessary. Figure 7.4 shows a

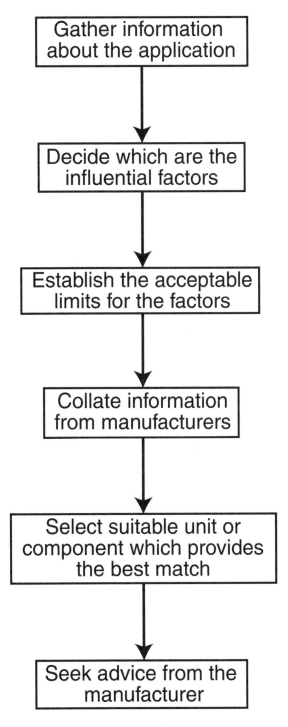

Figure 7.4 Unit or component selection procedure

generalized procedure for selecting the elements of a system. It shows the necessity for thorough information gathering regarding the application before selection can take place.

It is essential at the outset to define the boundaries within which the element must perform as comprehensively as possible. The information gathered relates to the purpose for which the element is required and to the criteria of life, performance, cost and operating environment with which it must comply. The information is needed in order to understand the total system so that the selection is consistent with the rest of the system. The temptation of the design engineer to consider only functionality should be resisted and time and effort expended at this stage is invariably rewarded when an optimum design is selected.

Factors which influence the selection of units and components must now be identified. The most important and common factors governing selection are often performance, application, geometry, environment, safety and commercial. Not all these factors are important in every design so careful study of the system is required to ensure that those considered are actually relevant. Reference should be made to the Product Design Specification for the system.

Each factor should be defined in terms which are as objective as possible. Thus, where appropriate and possible, numerical information should be given, terms must be explained and vagueness avoided. Following this the boundaries of satisfaction must be defined for each of the chosen factors. This assists the design engineer in making a selection which meets the stated requirements in every respect. Subjective judgements cannot always be avoided and in such cases a means of comparison must be established.

Manufacturers' data should be collated and arranged in a suitable format. There is a finite number of particular types of component available from manufacturers and the selection process is heavily constrained by the form and content of the information presented by them and the range of catalogues available to the design engineer when the need arises. There is a good case for maintaining a 'rolling' catalogue library, or data on microfilm/computer, since gathering such information can be very time consuming, particularly if a unique set of data is collected on each occasion. Data on size, cost and performance can often be noted in numerical form, giving a range where appropriate. In the case of less objective data a rating may be shown based on advice or opinion gathered.

Optimizing the choice is now a matter of selecting the best compromise, in the opinion of the design team, between the priorities of the system and the availability of hardware. In the initial stages some pruning of potential alternatives is called for. As far as the factors involving numerical data are concerned, some yield a go/no-go situation which will eliminate those which do not fit within the boundaries set.

Other requirements of a more subjective nature should be compared on the basis of the elements' ability to meet the criteria as laid down in the Product Design Specification. The evaluation technique used here is similar to that used elsewhere in the design activity, particularly for initial concept selection. References for further reading are included at the end of each subsequent chapter and many elaborate on the details of a variety of techniques.

Further advice on the detail of installation or specifying and ordering will be required from manufacturers' information. Normally this would be available in a catalogue but often it is necessary to communicate directly with a representative of the company.

As an example of a more detailed component selection procedure consider the decisions which influence the type of spring most suitable for a particular application. The flow chart in Fig. 7.5 illustrates this. The selection of a type of spring depends mainly on the space available and the magnitude and direction of the loading. One further important

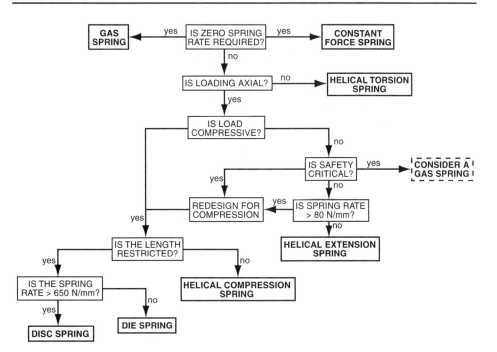

Figure 7.5 Selection chart for spring type

consideration which influences the selection of a particular type of spring is that for some applications safety codes require that a compression spring is used. This is because a failed compression spring can continue to provide a stop and hold components apart, in effect providing a fail safe design.

The flow chart in Fig. 7.5 illustrates the level of thought which must be applied to the smallest detail in any design. It only takes the failure of a relatively insignificant component to render the whole design worthless.

7.4 Robust design

Design for reliability

The reason for the steady increase in reliability engineering stems from the increasing awareness that the cost of ownership of a product or system comprises two components. The first is the capital cost and the second is the cost of operating, administering, maintaining and replacing the product or system. The second outlay, the running cost, can often exceed the capital cost and is a function of reliability. Indeed, because of the disastrous financial consequences of equipment failure most customers specify tightly reliability conditions.

One hundred per cent reliability testing is unthinkable since this implies that there would be no products for sale. The time required for reliability testing depends on the failure rate of the items under test. In general reliability adds cost to a product and although unreliability carries with it a cost penalty, the optimum level of reliability is always a

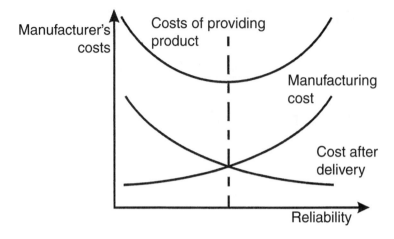

Figure 7.6 Cost of reliability

compromise between the two. Figure 7.6 shows the general relationship between reliability and cost.

Reliability is concerned with the causes, distribution and prediction of failure. Failure is defined as the termination of the ability of a component or system to perform its required function. The parameter, 'failure rate' is given the symbol $\lambda(t)$. Another method of describing the occurrence of failures is to state the mean time between successive failures. The two terms used, the mean time between failures (MTBF) and the mean time to fail (MTTF) are explained diagrammatically in Fig. 7.7. In many, but not all, cases MTTF and MTBF are the same. MTTF is the mean operating time between successive failures and the difference between the two terms is repair time. Hence,

$$\text{MTTF} + \text{mean time to repair} = \text{MTBF}$$

Components or systems which are not repaired do not extend beyond the point marked ▲ in which case MTTF and MTBF are the same.

The failure rate is not necessarily constant. If a reliability test were undertaken with a large sample and each product were tested until it failed and not replaced, the typical

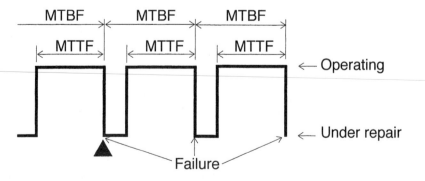

Figure 7.7 Difference between MTBF and MTFF

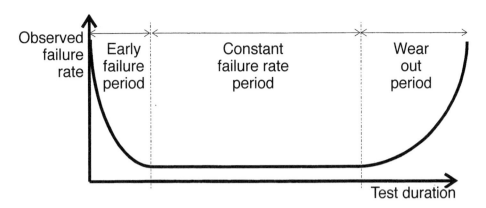

Figure 7.8 Failure rate against time curve

failure rate against time curve would be what is known as the 'bath-tub' curve shown in Fig. 7.8.

During the *early failure period*, within hours of commencement of operation it is quite likely that failures will occur due to various imperfections acquired during the manufacturing process, due to design faults or misuse. Gradually, these early failures will occur less frequently. This period is often covered by the manufacturer's guarantee.

The *constant failure rate period* is usually relatively long and the failure rate is normally approximately constant. During this period it is usual for failures to occur at a relatively low rate but from a wide variety of causes.

The beginning of the *wear out period* corresponds to the end of the useful working life. All products wear out and this occurs because of a variety of time-dependent mechanisms.

Reliability $R(t)$ is time related, and is defined as a probability and expressed as a value between 0 and 100%. Consider a number of components N_o being tested and allowed to fail without replacement so that at any time t there are N_s surviving. Then

$$R(t) = N_s(t)/N_o$$

If now the failure rate is expressed in terms of $N_s(t)$ the relationship between failure rate and $R(t)$ is

$$\lambda(t) = dN_s(t)/N_s(t) \, dt$$

Solving these two equations when the failure rate is constant gives

$$R(t) = e^{-\lambda t}$$

For this condition, the MTBF (θ) is the reciprocal of the failure rate, hence

$$R(t) = e^{-t/\theta}$$

As an example, records kept on 1000 engines of the same type show an average life to failure of 14 000 flying hours. What is the probability that one such engine will survive a transatlantic flight of 7 hours?

$$\text{MTBF} = 1/\lambda = 14\,000 \text{ hours}$$

$$\lambda = 1/14\,000$$

$$R(t) = e^{-\lambda t}$$

$R(7)$ = Probability of surviving 7 or more hours
$$= e^{-7/14\,000}$$
$$= 0.9995 \text{ or } \underline{99.95\%}$$

This indicates that engine survival is relatively assured. The situation is obviously more complex since a modern passenger aircraft will always have more than one engine and could continue flying without one engine. Hence, 99.95% is only a measure of the reliability of the engine not the aircraft. Since this is only an introduction to design for reliability, the use of distributions, such as the binomial, which would indicate overall aircraft safety based on system reliability, is not covered.

Whilst it is failure rate, MTBF, and hence reliability of components that is measured, it is the reliability of complete systems that is the ultimate concern of the designer, salesman and customer. The reliability of a system can be obtained from the knowledge of the reliabilities of its components. Whatever the system, failure of one component may cause the whole system to fail. For example, a domestic television may have 500 components, and manned spacecraft several million. Thus, the problem facing designers is not so much that the parts are unreliable but that there are so many of them. Many types of system–component relationships exist. Series and parallel are two such relationships.

Consider a system comprising two components connected in series such that failure of either causes system failure. The reliability of the system is the product of the component reliabilities. If the components each have a reliability of 90% then

$$\text{Reliability of the system} = 0.9 \times 0.9 = 0.81 \text{ or } 81\%$$

Expanding to the case of N components connected in series;

$$R_{an} = R_a.R_b \ldots R_n$$

R_{an} = reliability of the system

If a series system comprises 100 components each with a reliability of 90% then

$$R_{100} = (0.9)^{100} = 0.000\,026$$

A reliability of this order implies that there is no hope of the system working satisfactorily over the required lifetime. If the individual reliability is increased to 0.999 then the system reliability becomes 0.906 or 90.6%. These examples are artificial since:

- It is unlikely to be essential for every component to function correctly in order to ensure system success.
- Critical components can be duplicated so that if failure occurs there is a spare to take over – this is called redundancy.

If the failure rates λ_a and λ_b apply to the two-component system such that

$$R_a = e^{-\lambda_a t} \quad \text{and} \quad R_b = e^{-\lambda_b t}$$

Applying the series rule gives

$$R_{ab} = e^{-(\lambda a + \lambda b)t}$$

This shows that the system is a constant failure-rate system of failure rate $(\lambda a + \lambda b)$.

Consider now another type of system, again comprising two components but so connected that if one fails the system does not fail. This is a parallel system because the failure of a single component does not cause system failure. The probability of system success in this case is now the probability that either or both units succeed, hence

$$R_{ab} = R_a + R_b - R_a R_b$$

In general

$$R_{an} = 1 - (1 - R_a)(1 - R_b) \ldots (1 - R_n)$$

In order to achieve the degree of reliability required it may be necessary to duplicate components so that if one component fails there is another available to carry on working. The following are examples of this technique which is called *redundancy*:

- Altimeter in aircraft. One is insufficient in case of malfunction. Two would pose the problem of identifying which is giving the correct reading if they differed. Three are thus required. This is called active redundancy.
- In the operating theatres of hospitals. If the mains electricity fails provision is made for switching to an emergency supply. This is standby redundancy.
- The spokes of a traditional bicycle wheel illustrate another kind of redundancy. Several can break and the wheel will still function. This is called partial redundancy.

Consider the example of a gearbox containing six rolling element bearings. The L_{10} life of a bearing, which is quoted by a manufacturer, indicates the number of cycles after which 10% of the bearings would fail. That is, each bearing has a reliability of 0.9. The gearbox has six bearings and all must function for the gearbox to operate.

The gearbox reliability $= (0.9)^6 = 0.54$ or 54%

Considering a shaft requiring a bearing at each end. The reliability of the two bearings is 0.81 or 81%.

If this is unacceptable, two bearings could be used at each end of the shaft. The model now is as shown and the reliability is increased.

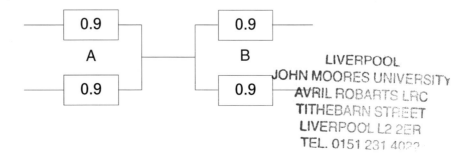

$$R_A = R_B = 1 - (1 - 0.9)^2$$

$$R_{shaft} = 0.99^2 = 0.98 \text{ or } 98\%$$

This gives an increase of 17% in reliability.

FMEA

Failure Modes and Effects Analysis (FMEA) was first used in the 1960s by the aerospace industry and is now a technique used by most industrial sectors. It is an objective method for evaluating system design. This is accomplished by establishing a multi-disciplinary team to consider all the potential failure modes of the components which make up a system or product and quantifying the influence such failures would have on overall reliability. It is one of the most powerful tools available for identifying reliability, safety, compliance, and product non-conformities during the design stages.

FMEA directs attention to those areas of a detail product design which may cause non-satisfaction of the reliability or safety criteria of the specification. Once critical components are identified then corrective action can be taken to improve the design. The technique can be used, for example, to identify critical parts within a system or product which benefit from the introduction of parallel or redundant components.

FMEA is an ongoing process that should start as a part of the first design review and continue throughout the life of the product. FMEA is a bottom up analysis technique.

The effect of a component failure depends upon the function of the component in the system. The severity of a potential failure is represented by the variable S and is assigned a value between 1 & 10, where 10 is the most severe. The occurrence of the failure (Relative Failure Rate) is represented by the variable O and is assigned a value between 1 & 10, where 10 is the highest failure rate. The ability to detect a failure is represented by the variable D which is assigned a value between 1 & 10 with 10 being the most difficult to detect. The relative importance of a failure mode is represented by its Risk Priority Number (RPN) calculated as

$$RPN = S \times O \times D$$

Every component has numerous potential failure modes and theoretically there is no limit to the depth one could go. Practically, there is a point of diminishing returns where the added cost exceeds the benefits derived. In practice a component with a RPN in excess of 100 is considered to be definitely worthy of attention. The FMEA process develops several very useful databases that provide manufacturers with the basic tools necessary to control the quality of their product.

Method
(1) A cross-functional team must be used to develop the FMEA.
(2) Identify the function of the component.
(3) List at least one potential failure mode for each function.
(4) Define the effects of failure in terms of what the customer might notice.

(5) Rate the severity (or seriousness) of the potential effect of the failure.
(6) Assign an occurrence rank to each of the potential causes/mechanisms of failure.
(7) Assign a detection ranking that assesses the ability of the design controls to detect a potential cause/mechanism or the ability of the design controls to detect the subsequent failure mode.
(8) Calculate the RPN numbers for each component.
(9) Identify and carry out remedial actions for potential significant and critical characteristics of components to lower the risk of the higher RPN failure modes.
(10) Calculate the new severity, occurrence, detection and RPN numbers.

The reasons for the FMEA being carried out by a multi-disciplinary team are explained by the subjective nature of arriving at the occurrence, severity and detectability scores. The following tables give general guidance for this process and the ratings quoted are those used in the automotive industry.

Table 7.3 Occurrence

Rating	Failure Rates	Probability of Failure
10	<1 in 2	Very High
8	1 in 8	High
6	1 in 80	Moderate
5	1 in 400	Occasional failures
3	1 in 15 000	Low
1	1 in 1 500 000	Remote: Failure is unlikely

Table 7.4 Severity

Rating	Effect	Severity of Effect
10	Hazardous – without warning	Potential failure mode affects safety or involves non-compliance with government regulation without warning.
9	Hazardous – with warning	Potential failure mode affects safety and/or involves non-compliance with government regulation with warning.
8	Very high	Item inoperable, with loss of primary function.
7	High	Item operable, but at reduced level of performance. Customer dissatisfied.
6	Moderate	Item operable, but customer experiences discomfort.
5	Low	Item operable, but at reduced level of performance. Customer experiences some dissatisfaction.
4	Very low	Poor fit or finish. Defect noticed by average customers.
3	minor	Poor fit or finish. Defect noticed by most customers.
2	Very minor	Poor fit or finish. Defect noticed by discerning customers.
1	None	No effect.

Table 7.5 Detectability

Rating	Detection	Criteria
10	Absolute uncertainty	Design Control will not detect a potential cause and subsequent failure mode.
9	Very remote	Very remote chance the Design Control will detect a potential cause and subsequent failure mode.
8	Remote	Remote chance Design Control will detect a potential cause and subsequent failure mode.
7	Very low	Very low chance Design Control will detect a potential cause and subsequent failure mode.
6	Low	Low chance Design Control will detect a potential cause and subsequent failure mode.
5	Moderate	Moderate chance Design Control will detect a potential cause and subsequent failure mode.
4	Moderately high	Moderately high chance Design Control will detect a potential cause and subsequent failure mode.
3	High	High chance Design Control will detect a potential cause and subsequent failure mode.
2	Very high	Very high chance Design Control will detect a potential cause and subsequent failure mode.
1	Almost certain	Design Controls will almost certainly detect a potential cause and subsequent failure mode.

Note: Zero (0) rankings for Severity, Occurrence or Detection are not allowed.

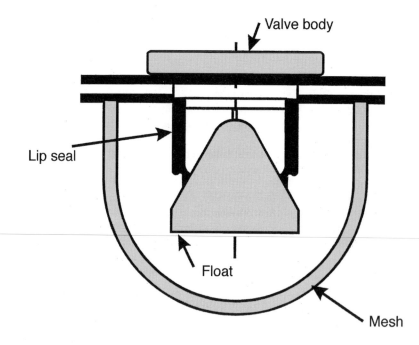

Figure 7.9 Pressure relief valve

Consider as an example the redesign of the pressure relief valve for a toilet cassette in a caravan. The redesign was required due to problems associated with 'blow-back' when the valve became blocked and therefore did not release pressure unless the toilet were operated. After consideration of various alternatives the concept shown in Fig. 7.9 was investigated using FMEA. The most important features of the new design aimed at increasing functionality are the rounded top of the float, preventing debris from settling, use of low friction coefficient material for the float and the mesh guard.

Table 7.6

Product function	Potential failure mode	Potential effects of failure	Potential causes of failure	Current controls	O	S	D	RPN
Equalize pressure between cassette and ambient	Valve fails to work	Cannot equalize pressure leakage	Mesh comes loose	Cleaning	4	4	3	48
			Valve blocks		5	5	2	50
			Float snaps off		2	5	2	20
			Seals perish		2	4	4	32
			Spring fails		1	5	2	10
			Plunger jams		1	5	3	15
			Valve works loose		1	4	2	8
			Chemical degradation		5	2	8	80
			Abuse in handling		6	4	1	24

Since there are no RPNs over 100 in Table 7.6 it is clear that there is no serious need for attention to any one element of the design. The two which are highlighted are the potential for chemical breakdown of the valve materials and possible clogging of the valve with solid waste. During detail design materials selection and surface finish were considered more carefully than might have been the case without carrying out an FMEA on the new design.

Design for quality

In the increasingly global competitive world in which companies must operate, persistent quality improvement, often allied with assembly and manufacture cost reduction, is essential if a company is to remain profitable. The quality of a product is a measure of the degree to which it meets the customer's requirements. Reliability is defined as the probability that a device or system will operate without failure for a given period of time. The difference between the two is the time element, reliability being concerned with how long quality exists.

A simple example which illustrates the difference between quality and reliability is a car tyre. The performance in cornering and braking deteriorates as the tyre wears. In other words, the quality deteriorates until it reaches an unacceptably low level and the tyre is replaced. By contrast, reliability is more a measure of the frequency with which tyre bursts occur.

The old concepts of improving product quality by controlling manufacturing processes and identifying poor quality items by inspection have given way to improving product and process design at the design stage. The main reasons for this dramatic switch of emphasis are illustrated in Fig. 4.2, which shows that the vast majority of engineering changes occur close to, and sometimes after, the product is released for production. The later in the design and development process these changes occur the greater the cost penalty. This is particularly true once manufacture has commenced and the effect is magnified by delaying the launch of the product. The aim is to design quality in by getting the product right first time. Expressed in terms of Fig. 4.2 the aim is to move the peak further left, as far before the release date as possible.

The adoption of the quality techniques which follow will result in shorter lead times, a reduced number of engineering changes, reduced costs and increased quality. Quality has no meaning unless it is related to cost, since, in general, the more expensive a product is the better its quality. The aim is to improve the quality of a product without increasing the cost of producing that product. This can be achieved by reducing the effects of variability of controlling parameters and the technique is called parameter design.

Within parameter design two different categories of factors are identified. The first category are control factors, which as the name suggests can be easily controlled. Examples of such factors include material selection, operating voltages and sizes. The second category are noise factors. These cannot be controlled easily and are often very costly when control is attempted. Examples include temperature and humidity. In an attempt at quantifying the effect of different parameters the signal to noise ratio has been introduced. This is so named because of the comparison in communications of the strength of a transmitted signal with the level of interference. There are three signal to noise ratios in general use: nominal, as in the case of colour density of a television; minimization, as for weight and noise; and maximization of such as strength or power.

The approach is to identify those control factors which are insensitive to noise factors and particularly those which exhibit non-linearity. The aim is to reduce the sensitivity of products to the source of variation. As an example consider the design of an electrical power supply circuit, the characteristics of which are illustrated in Fig. 7.10. Following conceptual (system) design the components making up the circuit are selected. There are many combinations of resistance, capacitance and transistor characteristics which will provide a functioning circuit.

If the desired output is 240 volts then parameter design seeks to reduce the variability about this mean for all circuits produced, without significant increase in associated costs. The output voltage can be determined by the gain on the transistor, which is non-linear. In order to obtain an output voltage of 240 V the designer selects gain A^1. If the actual gain deviation is as represented by the area under the normal curve then the voltage will also deviate about the target of 240 V. An alternative course of action is to use the effect of the non-linearity of the gain curve. If the flatter part of the curve were used, the target voltage would increase to around 270 V but the deviation about this value would be reduced.

Such adjustment of target value, and the introduction of an extra component, in this case a resistor to reduce 270 V to 240 V, must be considered carefully. It can often be a simpler and more cost effective way of controlling variance. Obviously the variance of the resistor must also be taken into account.

It is common practice to specify constraints in the PDS in terms of the limits of tolerance. If parameter design is to make its full impact then this must change and the constraints in the PDS must be specified in terms of target or ideal values with

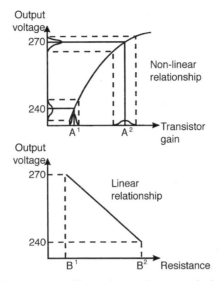

Figure 7.10 Transistor characteristics

accompanying limits of tolerance. Consider for example the fitting of a cab door. If the door is made to the smallest acceptable size and the frame to the largest acceptable size then the door will still operate satisfactorily but it may leak under certain conditions and allow more noise into the cab than a door and frame made to target. Target sizes guarantee optimum quality and should be the aim for the maximum amount of production possible.

Although it is difficult to quantify the quality of a product one measure is becoming accepted. This is called Loss to Society. The smaller the loss the greater the quality. A product causes losses when it deviates from target values, even when within the specified tolerance limits. An often quoted example, first published in a Japanese newspaper, involves two manufacturing plants making identical television sets for Sony, one in the USA and the other in Japan. The key characteristic involved was colour density, the target values for which were set after extensive customer trials.

The design philosophy in the Japanese plant was to aim for as many televisions as possible to be near to the target value for colour density and not to be too worried about a small number which were outside the acceptable limits. By contrast, the design philosophy of the American factory was to maximize the number within the acceptable limits. The results of the two philosophies are illustrated in Fig. 7.11. The Japanese philosophy resulted in a normal distribution whereas the American philosophy resulted in the flat profile. All of the televisions produced in the USA were within the specified limits of acceptability whilst a small percentage of the Japanese sets were outside these limits.

It is clear from the distributions in Fig. 7.11 that more of the Japanese sets were on and around the target value. Customer perception was that the Japanese sets were of much higher quality than the American sets. The conclusion that can be drawn is that quality is best achieved by minimizing variance rather than by strict adherence to the specification. The loss to society in this case is represented by a quadratic function with the loss to society increasing as the square of the deviation from target.

Of foremost importance is the definition of *'manufacturing' quality* as *product uniformity around a target* rather than conformance to specification limits.

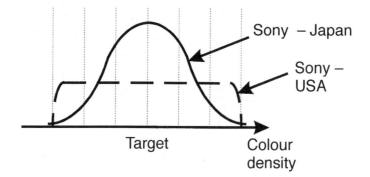

Figure 7.11 Colour density philosophies

Measures to counteract and reduce product variability can only be taken during design. Techniques used later, during manufacturing or process design can reduce variance introduced due to, for example, manufacturing tolerances and material imperfections. However, even these can be most effectively tackled during design.

Design against high cycle fatigue

Only rarely does the failure of a component or structure occur due to the application of a single static heavy load. Failure is more often caused by lighter repeated or cyclic loading and usually occurs at a point of stress concentration where the shape changes abruptly. The mechanisms by which cracks are propagated through the material are generally of more interest to the materials scientist than the designer. What is important for the designer is to recognize the features which influence fatigue life and to modify the design accordingly. Some very simple rules can be applied which will extend the life of a product greatly.

- Reduce stress raising effects.
- Provide best possible surface finish.
- Compress the surface.
- Stress relieve.
- Select materials which resist fatigue.

High cycle fatigue, above 1000 cycles during expected design life, is by far the most common cause of component failure in service accounting for an estimated 80% of all fractures. In the vast majority of cases the cause of the failure can be traced back to a lack of detail design consideration, particularly in component or joint shape. This is because most failures occur at changes in cross-section of a component or at the joints between components. Where this change in shape is accompanied by the welding of two components together the potential for disaster is significantly increased. This is due in the main to residual tensile stresses caused by the cooling and contracting of the weld pool. There are many factors which affect fatigue life.

Fortunately there are some general guiding principles and analogies which, if followed, will result in improved fatigue life. The main analogy used for predicting highly stressed regions of components and joints and therefore likely points at which fatigue failure will

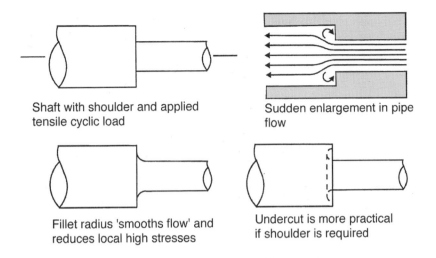

Shaft with shoulder and applied tensile cyclic load

Sudden enlargement in pipe flow

Fillet radius 'smooths flow' and reduces local high stresses

Undercut is more practical if shoulder is required

Figure 7.12 Stress and fluid flow analogy

occur is to compare the structure with fluid flow through pipes or channels. Consider the simple comparison, shown in Fig. 7.12, of a piece of round bar with a shoulder loaded in tension with pipe flow through a sudden expansion. The load is the same along the length of the bar, so where the cross-sectional area is large the stress will be low and the stress will be relatively large when the area is small. In the pipe flow analogy the amount of water flowing in each section is constant so the flow rate must increase in smaller sections. Two other features of pipe flow are helpful. Where eddy currents exist this indicates a region of very low stress and where the flow lines are locally closer there will be a local increase in stress.

Thus stress and velocity are related in the simple case of a steady flow. Perhaps surprisingly, the analogy also holds true where sudden changes in shape or flow restrictions cause turbulence. At the shoulder in the round bar we get a stress concentration whilst in the pipe the flow at the sharp corner is much faster. In the pipe flow there is also an area of slack water where the main flow cannot follow the walls into the corner. In the same position of the bar there will be a 'stress shadow'. This is a region of relatively low stress which could be used to advantage.

If the sharp corners were blended or radii then the local disturbance to the flow would be reduced and the highest levels of stress reduced significantly. In the pipe flow this would allow the streamlines to separate or come closer together gradually. The design of the shoulder on the shaft can be improved to 'smooth the flow' and reduce localized stress concentrations by the addition of a corner radius. However, if a shoulder is required for positioning of a bearing axially on the shaft then the undercut shown in the fourth sketch is probably the best option. Notice that the fatigue life is being improved by the removal of material.

Using the pipe flow analogy it becomes relatively easy to predict areas of high stress and therefore areas at high risk of fatigue failure. The technique, illustrated in the diagrams of Fig. 7.13 is to draw out a cross-section and then to rub out all internal outlines, leaving the shape of the metal. You now imagine that the section has water flowing through it along the direction of the applied load. In the first of the diagrams a structure to be loaded in tension has four fillet welds. In the second diagram the weld material is assumed to be

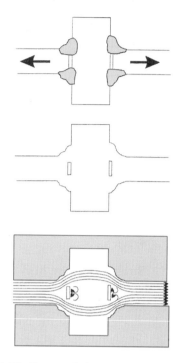

Figure 7.13 Determining high stress regions

homogeneous with the original metal. As illustrated in the final diagram the pipe flow analogy shows two potential areas of high stress. These are caused by the lack of penetration of the weld causing a blockage in the centre of the flow and the corners of the weld being sharp at the point of expansion and contraction of the flow.

It is already established that a sudden notch or change in section will cause local stresses to be significantly increased. Extending the flow analogy a little further, it is at first sight surprising that removal of material to give a smooth profile can also extend fatigue life. Figure 7.14 shows a few examples of poor weld shapes which could be filled out by extra weld or could be dealt with more effectively by grinding away the sudden change in section.

Many manufacturing processes leave residual or built in stresses in the component or structure. Most common of these are tensile stresses due to differential cooling following welding. These will relax with time but there are various techniques for speeding the process. Shot blasting, sand blasting or any similar process which compresses the surface reduces significantly any built in tensile stresses. In stress relieving the structure is heated and then allowed to relax.

Imagine a small crack in a component of width 4 µm. The fatigue damage from each cycle will be proportional to crack movement, so a tensile load which opens the crack to 8 µm before allowing the component to relax does 4 µm worth of damage. In the same way, a compressive load closing the crack from 4 µm will do 4 µm worth of damage. A tension followed by a compression would do 8 µm worth of damage. Heating the structure to around 650°C followed by a slow cooling allows local high stresses to even out and cracks to reduce in width as a result. Following stress relieving, tensile loads still have the same

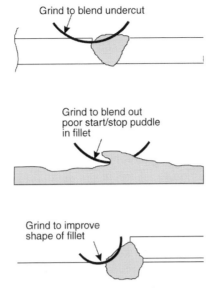

Grind to blend undercut

Grind to blend out
poor start/stop puddle
in fillet

Grind to improve
shape of fillet

Figure 7.14 Removing material improves strength

damaging effect. However, since the crack is now closed up the damage caused by compression is reduced by approximately 75%.

There is a great deal of very advanced science in the most refined areas of fatigue life prediction, but a good deal of common sense and the application of these basic principles should ensure that many potential failures are avoided. Materials which exhibit a greater resistance to fatigue should be selected for critical areas of a design. For example, ceramic materials are not thought to be susceptible to fatigue at all. Although general awareness of the phenomenon of fatigue began with the Comet aircraft disasters some years ago, the underlying principles were appreciated more than a century ago by the British engineer Sir William Fairbairn, who carried out classic experiments on wrought-iron girders. He found that a girder which was statically loaded would support 12 tonf for an indefinite period but would fail if a load of only 3 tonf were raised and lowered on it more than a million times.

The quantification of the number of cycles that a product will last for, fatigue life, is based on testing data. There is a great wealth of this data which it is beyond the scope of this text to reproduce. However, the factors which must be taken into account in design are:

- mean stress
- alternating stress
- material ultimate tensile strength
- the type of loading: bending, axial or torsion
- the surface finish (the rougher the surface the lower the number of cycles the component will complete)
- the effect of any stress raiser
- the size of the component (since only relatively small components are normally tested to destruction).

7.5 Principles

Detail design principles

Optimization   The search is for the best compromise between conflicting criteria.

Simplification   Where possible rely on the expertise and knowledge of specialist manufacturers by using proprietary components.

Analysis   Ensure that all components have appropriate factors of safety and are not over designed.

Robustness   The designer should aim for a product which is fit for the purpose intended for the lifetime intended.

Synthesis   A solution is often arrived at by a combination of techniques and elements.

Iteration   Progress towards the production stage is made iteratively as knowledge of the important factors grows.

7.6 Exercises

1. Bolts installed on a production line are tightened with automatic wrenches. They are to be tightened sufficiently to yield the full cross-section in order to produce the highest possible initial tension. The limiting condition is twisting off the bolt head during assembly. The bolts have a mean twisting off torque of 20 N m with a standard deviation of 1 N m. The automatic wrenches have a standard deviation of 1.5 N m. What mean value of torque wrench setting would result in only 1 in 500 twisting off during assembly?

2. A shaft tolerance has a standard deviation of 0.01 mm. The hole tolerance has a standard deviation of 0.016 mm. The difference between the means is 0.045 mm. If the entire production is accepted for assembly, determine the proportion of assemblies with clearance less than the allowance of 0.01 mm and the proportion of assemblies expected to interfere.

3. A unit has a constant failure rate of 0.3% per 1000 hours. What is its MTBF? What are the probabilities of the unit successfully completing missions of 10 000, 100 000 and 1 000 000 hours?

8 Design management

Two different levels of design project management are presented, those most appropriate for student project work and those more advanced methods employed by companies. The relatively new International Standards covering design for quality are explained and the engineering design management control process outlined. Techniques covered include project planning and control by means of bar charts and Programme Evaluation and Review Technique (PERT) network analysis. An introduction to Quality Function Deployment (QFD) is presented, the requirements for formal design reviews are explained and the Value Analysis technique discussed.

8.1 Introduction

As Chapter 1 illustrates we have been designing products for many centuries. Despite this the standards of many modern products in terms of design leave much to be desired. Most people have experienced products which are ugly, difficult to use, unreliable, difficult to maintain, too costly or not intrinsically safe. It is also true that even today many products are very difficult to dispose of at the end of their useful life even though design for recycling is acknowledged as of paramount importance. A recent survey conducted by the UK Design Council concluded that an average product could be redesigned to reduce manufacturing costs by 24% and to improve market demand by 29%.

The message is that design must be managed more effectively. The assertion that design is a creative activity and cannot be managed is not true. Reference to the design process shown in Fig. 1.5 illustrates that creativity is just one section of the design process as a whole and that much of what is called design can and must be controlled. Much more complex tasks, such as planning, opening and running a new factory or developing a new market, are successfully managed. It is essential that companies manage the design process and it is equally essential that students manage project work with the same rigour.

It is also important to manage the design function since, as illustrated in Fig. 3.1, the cost of manufacturing a product is mainly dictated by design. Many enlightened companies recognize this fact and are placing much greater emphasis on the early stages of the design process. Upstream design changes do not cost much. As we proceed downstream we find that the cost per phase increases exponentially.

The prime, and some would say only, reason for a manufacturing company to exist is to make profits for shareholders. If a company does not achieve this then it will go out of business. The design function has a major role to play in this profit making and competitive and therefore successful products are essential. It is possible to make enormous profits from single innovative ideas for a short time, without a professional and comprehensive design approach. However, if a company wishes to stay competitive over a period of time then a well managed, resourced and modern approach to design is essential.

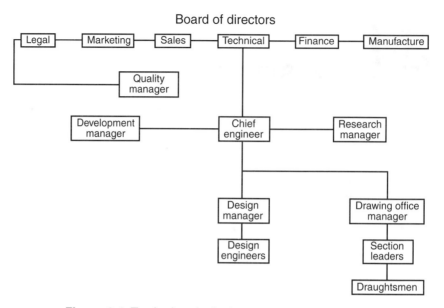

Figure 8.1 Typical technical management structure

A necessary precursor to successful management of design is that the company wide management structure be clearly defined. The company wide interfaces with design were discussed in Chapter 1 and illustrated in Fig. 1.6. It is clear that the design department must communicate with many other departments within a company on a regular basis. It is important also to consider the typical management structure as illustrated in Fig. 8.1. The major divisions are generally sales, manufacturing and technical. The chief engineer is normally responsible to the technical director and is responsible for research, development and design. Design engineers generally work in teams and the major output is data, in the form of scheme drawings and supporting documentation, which is fed via the drawing office manager to draughtsmen.

8.2 Management of design for quality

In recent years there has been a recognition that improvements in the quality control of the design process itself were necessary. Figure 8.2 illustrates the major stages of the design process with the necessary management control elements. These elements are project planning, Quality Function Deployment (QFD), design reviews and value analysis/engineering. QFD is a series of management control matrices which helps keep customer requirements firmly to the fore during all stages of a particular project. Each of these management control elements is covered in more detail later in the chapter.

The management of design for quality is governed by standards with which an accredited company must comply. Most countries have adopted the international standard whose constituent parts have collectively become known as ISO 9000 and include ISO 9000–9004. ISO 9000 is a guide to the selection and use of the appropriate part of the ISO 9000 series.

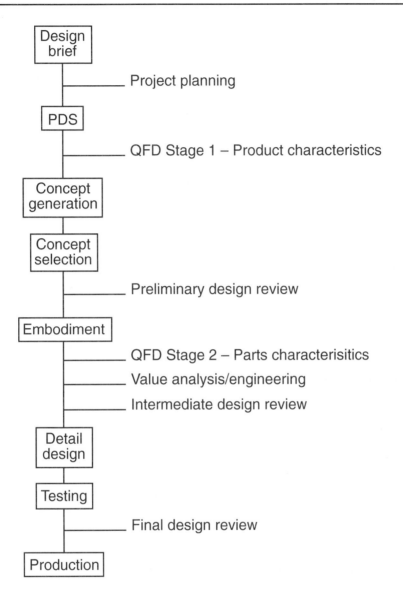

Figure 8.2 Engineering design management process

ISO 9001 relates to quality specifications for design, development, production, installation, and servicing when the requirements of goods or services are specified by the customer. ISO 9002 sets out requirements where a firm is manufacturing goods or offering a service to a published specification or to the customer's specification. ISO 9003 specifies the quality system to be used in final inspection and test procedures. ISO 9004 is a guide to overall quality management and the quality system elements within the ISO 9000 series.

Collectively these standards set out how a company can establish, document and maintain an effective quality management system which demonstrates to customers a commitment to quality. The standards cover the whole range of company activities and the management of design is (in the main) covered by ISO 9000 and 9001. The needs are to establish and control the functions of design planning, assigning activities to qualified staff with adequate resources, controlling interfaces between different disciplines, documenting design input requirements and design output in terms of requirements and calculations. Design output must be assessed to verify that it meets design input requirements and documented procedures must be used to control all design changes and modifications. Careful, planned and documented control at each stage ensures a smooth passage from concept to end product.

There are many stages and methods recommended by quality standards for the effective control of the design process. These are simplified and represented in a form which could be used in student project work.

(1) A design programme which breaks down the design process into separate elements and is presented as a chart of the design activities against time. The chart should show major events such as design reviews so that progress can be measured.

(2) Design management control procedures should be documented to show the relationship with various departments and with contractors. It is essential that the design team is not isolated from staff of other disciplines. Design responsibilities should be defined at a level appropriate to the design task. Design documentation should consist of numbered drawings and specifications, log books, calculations with assumptions, sketches, the design concept, analysis and test results.

(3) If innovation is proposed, then the desirability of each innovation should be assessed by analysis and/or testing in order to justify objectively its adoption in preference to established alternatives. For the introduction of a new material, tests must be conducted and results compared with the stated performance specification.

(4) Procedures should be prescribed for the identification and revision status of design documents, records of changes made and their distribution, control, recall and for the approval of all document changes by the person responsible for the design.

(5) Design control should ensure that tolerances are adequate to provide the required quality. They should be no tighter than necessary.

(6) National and international legal requirements, including health and safety standards, placing constraints on designs should be identified.

(7) Reliability requirements should be included in the specification. The specification should contain details of the proposed reliability testing programme and how the test results are to be analysed.

(8) Value engineering tasks should be undertaken and documented.

(9) Systematic procedures should be established to ensure the use of data gained from previous designs and user experience.

(10) Regular design reviews, which are documented and systematic critical appraisals of the design, must take place. The objectives are to ensure that the design satisfies the specification, that other viable solutions have been considered and that the design can be produced, inspected, installed, operated and maintained in a satisfactory manner. The design reviews must also ensure that there is adequate supporting documentation defining the design.

It is obviously impossible outside a company environment and particularly during a student project for all of these factors to be taken into account. However, some form of control of the design process is always desirable. The steps recommended for the management of a complete engineering design project are listed in Fig. 8.2.

8.3 Project planning and control

The two primary tasks involved in project management are planning and control. During the planning phase project timescales, costs and resources must be identified. During the project the emphasis is on the use of simple methods which will readily provide key information for the control of the project without creating too many demands on time and resources.

Gantt chart

A simple bar chart, as illustrated in Fig. 8.3, is probably the most widely used technique for monitoring of projects. Bar charts can be developed for a whole project, as illustrated, or can be used to identify the effort required by particular personnel at particular times. The chart in Fig. 8.3 is unusual since milestones, in the shape of design reviews, are indicated. Also, in the industrial environment it is more usual to employ months as the units of time. However, student-based project work is of necessity generally of relatively short duration and weeks are more appropriate.

The horizontal bars on the chart are all blank at the start of a project and are filled in as work progresses. In Fig. 8.3 the project has reached the end of week 6. The chart indicates that the embodiment work is ahead of schedule but that the cost estimate for the selected concept is not complete. This could be critical since the careful management of costs must run in parallel with the management of time and decisions made during the preliminary design review are based to some extent on cost information. In short the chart indicates where efforts need to be increased if the original targets are to be met.

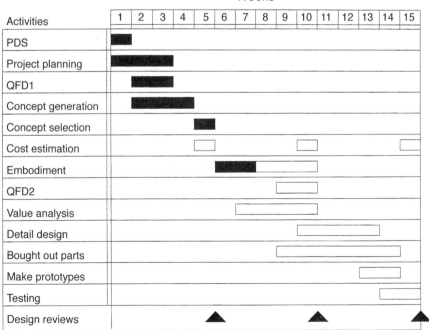

Figure 8.3 Bar chart for design project

It is essential that for all project work, particularly student design project work, that as detailed a bar chart as possible is developed at the outset. It then follows that the chart should be regularly updated and the bars filled in so that progress can be monitored.

It is worth noting that one of the main aims in project planning is to reduce lead times to the minimum possible so that a product may be on sale and earning profits at the earliest opportunity. This is why there should be a significant amount of overlap of the activities on any bar chart. One disadvantage of the bar chart is that the complex interdependence of activities and the dependence of one activity on the completion of another is not clearly indicated.

PERT networks

There are many network techniques and the Critical Path Method (CPM) and Programme Review and Evaluation Technique (PERT) are the most appropriate for design project management. CPM and PERT do not differ greatly but since PERT is more widely employed in industry and is suited to projects with well defined goals it is presented here.

A simplified PERT network analysis diagram for the off-highway vehicle seat suspension system is shown in Fig. 8.4. The numbers in circles represent completed *events* where 01 is the start event and 20, final design review, is the end event. The arrows indicate *activities* and only one arrow can connect any two events. Activity times in weeks are shown at the head of each arrow. In a complete PERT analysis, optimistic and pessimistic forecasts for the duration of each activity are included and distribution analyses performed, both of which are beyond the scope of this book.

There are many rules which govern PERT networks and some of these lead to the existence of *dummy activities*. It is sufficient to say here that these absorb no resources and are used to clarify the network and to get round layout difficulties which could otherwise

Table 8.1 PERT network events table

Label	Event
01	Project start
02	PDS fully defined
03	Project planning complete
04	QFD Stage 1 complete
05	Concept generation complete
06	Concept selection complete
07	Cost estimate of selected concept complete
08	Preliminary design review
09	Embodiment partially complete
10	QFD Stage 2 complete
11	Value Analysis/Engineering complete
12	Cost estimate of embodied design complete
13	Embodiment complete
14	Intermediate design review
15	Detail design complete
16	Bought out components ordered
17	Cost estimate of detail design complete
18	Prototypes manufactured and assembled
19	Testing complete
20	Final design review

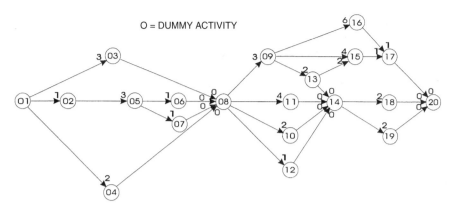

Figure 8.4 PERT network for suspension system

arise. These can either be represented by dotted lines or by the 0s used in Fig. 8.4. To facilitate better understanding it is normal to accompany the network with two tables, an event table (Table 8.1) and an activity table (Table 8.2).

Table 8.2 PERT network activity table

Label	Activity	Weeks
01–02	Define the PDS	1
01–03	Define the project plan	3
01–04	Define QFD Stage 1 matrix	2
02–05	Generate concepts	3
03–08	Dummy	
04–08	Dummy	
05–06	Concept selection	1
05–07	Estimate cost of concept	1
06–08	Dummy	
07–08	Dummy	
08–09	Start embodiment	3
08–10	Define QFD Stage 2 matrix	2
08–11	Value analysis/engineering	4
08–12	Estimate cost of embodiment	1
09–13	Finish embodiment	2
09–15	Start detail design	4
09–16	Order bought out components	6
10–14	Dummy	
11–14	Dummy	
12–14	Dummy	
13–14	Dummy	
13–15	Finish detail design	2
14–18	Manufacture of prototypes	2
14–19	Testing	2
15–17	Estimate cost of detail design	1
16–17	Estimate cost of bought out parts	1
17–20	Dummy	
18–20	Dummy	
19–20	Dummy	

Once the network is verified and the estimated activity times are available, the network analysis can begin. The simplest task is to identify the critical path, which in this case is 01.02.05.06.08.09.16.17.20 and is fifteen weeks in total. If during the course of a project any activity on this critical path takes longer than expected then the total project time will be increased. All other paths are considered non-critical and therefore have *slack* time which allows flexibility in the start and finish times of the activities.

As mentioned earlier, probability analyses can be performed on PERT networks and figures indicating the level of certainty with which the overall project time can be established can be computed.

Delta analysis

This is a technique favoured by managers of projects which are generally not based in engineering. Why the technique does not find favour in engineering is unclear since it is a very simple and visual concept to grasp.

During a project there are three parameters which are monitored, technical progress, cost and time. In Delta analysis the difficulty of monitoring technical progress is overcome by measuring it in terms of time and cost. In Delta analysis project progress is measured as a percentage of objective achievement and the absolute units of progress are irrelevant. This is illustrated in the top part of Fig. 8.5 where there is a progress shortfall.

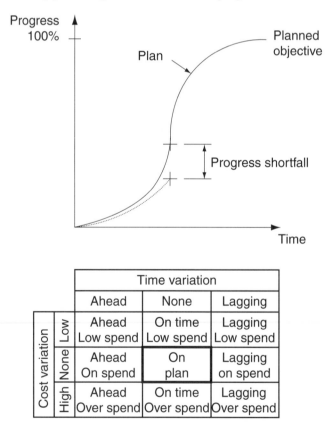

Figure 8.5 Project control – time and cost

The plan includes both cost and time targets and, as indicated in the lower part of Fig. 8.5, nine possible combinations of levels of progress can exist. Obviously it is relatively easy to identify within which box a particular project lies at any one time. The difficulty comes in the quantification of the level of variation.

8.4 Product Design Specification (PDS)

Although not identified as a formal management tool in Fig. 8.2 the importance of the PDS cannot be overstated. This is a formal document interfacing between marketing and engineering functions. The aim is to convert the perceived market need into functions and constraints covering the products' design, manufacture and marketability. This is the seminal document on which the whole design hinges. The development of a PDS is covered in detail in Chapter 2.

8.5 Quality Function Deployment (QFD)

What follows is a simplified introduction to QFD. The prime objective for a manufacturer is to bring a product to the market sooner than the competition with lower costs and/or better quality. One of the means of achieving this is QFD. This is a process whereby customer needs, wants and values are translated into the appropriate technical requirements for each stage of the project. All too often engineering designers imagine that they fully understand the customer's requirements or that they know better. Clearly such situations must be avoided if a competitive product is to be produced.

By considering the voice of the customer and carrying this through all stages of the product life cycle a continuous framework for product development is provided. QFD is a design methodology targeting engineering expertise and professionalism towards satisfying the needs of the customer, exposing in the process unrecognized aspects of the customer's expectations and identifying the necessary actions to satisfy them.

Whilst the ultimate end customer for a product is relatively easily identified, many other intermediate customers are likely to be found. Each intermediate customer will have a separate and distinct set of expectations which need to be taken into account. The challenge for the designer of a product is to satisfy these customer needs in the most cost effective way whilst optimizing requirements which often conflict.

The two major processes in QFD are product quality deployment, which translates customer language into technical specifications and deployment of the quality function, the definition of the overall manufacturing process. This is a method used to control the whole design and manufacture process. QFD provides a framework for structured and systematic communication between marketing, engineering and manufacturing and facilitates simultaneous engineering.

The full QFD cascade covers the complete project and four matrices are developed at different stages. As illustrated in Fig. 8.6 the first of these is the customer requirements against design requirements matrix where key product characteristics are determined and it is developed at the design concept stage. In the second QFD stage design requirements are compared with component characteristics. The third matrix compares component and process characteristics and during the fourth stage process characteristics are compared with production operations. The first two stages only are employed during the design process.

Many advantages can be claimed for QFD and these include:

- QFD is simple and structured;
- QFD helps to focus decisions;
- QFD provides traceability of decisions;
- QFD provides a common format for the whole project.

A QFD analysis begins with the establishment of who the customers are and the identification of their needs. The customers may be the end user, the production department who will have to manufacture and assemble the device or the sales and marketing departments. For each tertiary customer requirement identified the method of achieving it will create a list of design attributes. What follows are some of the end user customer requirements, reflecting the often imprecise nature of such requirements, for the seat suspension mechanism introduced in earlier chapters.

(1) The adjustment of the seat position should be easy to accomplish. It should be possible to do this quickly.
(2) The operator must remain comfortable and work to maximum effect for an eight hour shift.
(3) The operator must be safe.
(4) Installation of the seat and suspension system should be easy and quick.
(5) The combination of the seat and suspension system should be floor mounted.
(6) The cost of the suspension system should be the minimum possible.

This list is incomplete and presented for illustrative purposes only. The development of a full list of customer requirements is achieved iteratively and is guided by the use of a check

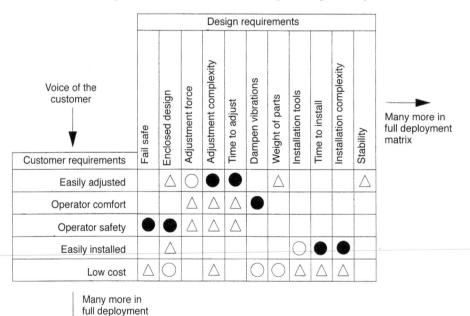

Figure 8.6 QFD Stage 1 – customer and design requirements. △: weak relationship, ○: medium relationship, ●: strong relationship

list such as in Figs 2.5 and 9.2. These customer requirements must be converted into design requirements which are measurable. The list of design requirements cannot be divorced from the PDS and must be agreed with the customer before the design process can proceed.

To facilitate interpretation, a matrix is drawn with customer requirements listed down the side and design requirements across the top. Using such a matrix, Fig. 8.6 shows part of the matrix developed for the seat suspension, indicates whether customer requirements are adequately represented in the design requirements. For example, since the low cost customer requirement has only relatively weak relationship signs the indication is that the issue of cost is not adequately covered by the design requirements. The matrix can be further used to establish a customer importance rating for each design attribute by assigning values to each of the signs (say 1, 3 and 9) and totalling each column or design attribute. Obviously the design team should concentrate their efforts on those design requirements which have the highest ratings.

At this point it should be noted that a perceived shortfall in the usefulness of QFD is the lack of a concept selection stage. To this end enhanced QFD is being developed which will include similar procedures to those outlined in Chapter 4.

The design requirements dictate the design attributes and therefore the final product characteristics. Stage 2 of QFD, which is a matrix of design requirements and attributes, is illustrated in Fig. 8.7. This stage of the deployment indicates the extent of the critical

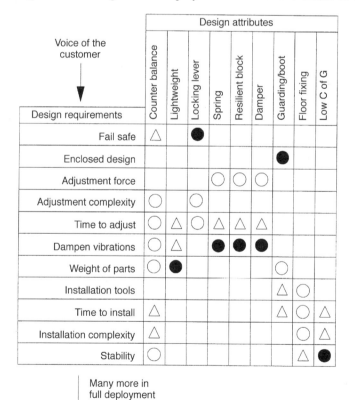

Design requirements	Counter balance	Lightweight	Locking lever	Spring	Resilient block	Damper	Guarding/boot	Floor fixing	Low C of G
Fail safe	△		●						
Enclosed design							●		
Adjustment force				○	○	○			
Adjustment complexity	○		○						
Time to adjust	○	△	○	△	△	△			
Dampen vibrations	○	△		●	●	●			
Weight of parts	○	●					○		
Installation tools							△	○	
Time to install	△						△	○	△
Installation complexity	△							○	△
Stability	○							△	●

Many more in full deployment matrix

Figure 8.7 QFD Stage 2 – design requirements and attributes. △: weak relationship, ○: medium relationship, ●: strong relationship

linking between the design requirements and the product characteristics, in this case for the seat suspension project. A fully developed list of product characteristics is deployed further to guide production planning and control.

This diagrammatic representation of the product features and design attributes is a powerful tool in a team working for identifying design trade-offs and critical features. It establishes a check list of relationships to be referred to at each stage of the design process. This matrix concept is developed to include importance rankings, the strength of the relationship, competitor ratings and customer ratings. The approach focuses attention on critical and novel aspects of the design.

8.6 Design review

Periodic review or audit of the design is vital since there is an overlap between innovation and practical skills and due to the fact that few design teams are made up of experts covering everything pertinent to the project. The design review is the crucial link between relevant experts and the design team.

The aims of the design review are to confirm the design meets the PDS, to ensure that the design can be produced economically and to highlight potential problems. The review team should consist of internal management, manufacturing, purchasing, marketing personnel and outside people, including customers and specialists as deemed necessary. The review team is organic in that the constitution should be varied depending upon the project. The intention is to appraise the design and not the designer! The design review provides a formal opportunity to assess progress, an early opportunity for customers to familiarize themselves with new equipment and provides the opportunity for the designer to be quizzed by people who would otherwise not be involved.

The aims of the three review periods indicated earlier are evaluation of specifications, analysis of potential failure modes, discussion of test philosophy, identification of potential problems for manufacture and inspection and checking drawings for ambiguity.

Preliminary review

The preliminary review should include:

(1) Checking of the PDS.
(2) Consideration of the generated concepts.
(3) Verification of the selection of an optimum concept.
(4) Consideration of research programmes.
(5) The distribution of the work between internal sections and subcontractors.
(6) The environmental conditions in which the product will be used.
(7) The method of use of the product.
(8) The proposed maintenance policy.
(9) The suitability of the concept for manufacture and assembly.
(10) Value engineering constraints, such as standardization and interchangeability.
(11) Plans for testing and evaluation of results.
(12) Preliminary cost estimates.
(13) Timescale estimates.

Intermediate review

In a complex project there should be many intermediate reviews and the design being studied should be split down to sub-assembly or component level. A separate review takes place on each and includes a reappraisal of all the considerations in the preliminary review and:

(1) Failure modes effects analysis.
(2) Stress analysis.
(3) Quality and reliability assessment.
(4) Inspection methods.
(5) Packaging and shipment.
(6) Operating instructions.

Final review

The final review should include all of those issues raised during earlier reviews and:

(1) The suitability of the design for manufacture and assembly.
(2) Cost and timescale estimates, including life cycle costing.
(3) Operating instructions and handbooks.
(4) Post design services.

During a review meeting full documentation of the major decisions, along with justification for those decisions, is essential. Only following the meeting and once the design team has reviewed the documentation from the meeting can the ultimate design configuration be established.

8.7 Value analysis/engineering

Value analysis is a technique which should be applied to all new designs. It involves questioning everything associated with the design and manufacture of a product with the aim of improving the value of a product. The essential difference between conventional cost cutting and value analysis is broadly that it involves reducing the cost and/or improving the functionality of the product.

In many companies value analysis was first employed by production engineers to reduce the cost of manufacture. However, it was quickly recognized that the impact of value analysis would be much greater if a multi-disciplinary team of engineers were formed which would also influence the design team. The ideal value analysis team should consist of personnel from design, purchasing, marketing, accounts and manufacturing. In modern terminology these teams are often referred to as multi-disciplinary Product Introduction Teams or Concurrent/Simultaneous Engineering teams.

During value analysis the major steps are:

(1) Definition of function

The analysis and definition of the function(s) which the product must perform. All functions should be identified and defined in two words. For example, the functions for our suspension example could be reduce vibration, enable adjustment and enable rotation. It should be noticed that each function consists of a verb and a (measurable) noun. Functions must be further classified into primary and secondary functions where the primary function is the main reason for the product.

In our case the primary function is to reduce vibration. In value analysis, one technique is to identify the cost of providing each function and, for example, if a secondary function is costing more than a primary function then something is clearly wrong with the product. This procedure often triggers ideas and information which result in improved value.

(2) Speculation on alternatives

Full information, such as that discussed in Chapter 9, must be available to the value analysis team before alternatives are created. These alternatives can be created at any level, including product, assembly or component levels. A series of questions, such as those listed and broken down into function, material, geometry and manufacture headings, serve to trigger the creative process.

Function

- Are all the secondary functions necessary?
- Is there anything else which will perform the function?
- Can any functions be combined?

Geometry

- Can the size be reduced?
- Can the weight be reduced?
- Are specified tolerances appropriate?

Manufacture

- Can assembly times and operations be reduced?
- Do standard jigs and fixtures exist?
- Are all operations necessary?
- Can operations be combined?
- Can operations be simplified?
- Could a standard or bought out part be used?
- Can machining and waste material be reduced?

Materials

- Can another material be used?
- Will a change of material influence size?
- Would a different material simplify manufacture?
- Is the proposed finish essential?
- Is it possible to use pre-finished materials?

These questions and others force the value engineer into accomplishing one or all of the following aims of value analysis:

- Reduction of the number of parts
- Reduction of the number of manufacturing operations
- Reduction of the complexity of manufacturing operations
- Introduction of alternative materials
- Use of standard or bought out parts
- Elimination of redundant features
- Relaxation of specified tolerances
- Use of pre-finished materials
- Rationalization of product ranges
- Reduction of machining and waste material

Some of these aims have been achieved in the simple bolted connection presented as Fig. 8.8. The detail design as originally intended had two more parts and a longer bolt than the value engineered design, in which the adapter plates have been joggled. Clearly, the value engineered design involves more work in the bending of the plates. On analysis it was found that the value engineered design reduced costs by 15%, primarily because of time saved on assembly, and that functionality was not affected.

(3) Evaluation and verification of alternatives

The economic feasibility of the created alternatives must be established and one or more selected for investigation of technical feasibility. Proof is required that the selected alternatives will cost less, be easier to maintain or will increase performance and reliability.

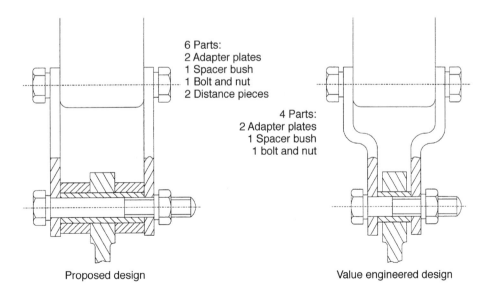

6 Parts:
2 Adapter plates
1 Spacer bush
1 Bolt and nut
2 Distance pieces

4 Parts:
2 Adapter plates
1 Spacer bush
1 bolt and nut

Proposed design

Value engineered design

Figure 8.8 Value engineered connection

(4) Presentation of recommendations

The preparation of a clear and concise formal proposal for the modification of the design and/or manufacture process is essential. All the advantages of the changes suggested must be quantified. If the report stage is handled in an insensitive way the design and production engineers whose work has been superseded by the value analysis team could be alienated.

8.8 Principles

Engineering design management principles

Planning   A clear plan must be established at the outset and subsequently used to guide each stage of the design process.

Control   It is essential that the design project is controlled by various means in order to avoid lack of direction and unsuitable products.

Value   The aim is to provide a solution within which the design attributes reflect the value of the customer.

Quality   Both the quality of the product, as judged by the customer and the quality procedures controlling its design must be the highest possible.

Review   At least three design reviews or audits are required as part of the design process. During these audits it is necessary to question whether any more resources should be devoted to the project, based upon how well the design is progressing when judged against the specification.

Documentation   At all stages decisions made and calculations performed along with geometrical information should be recorded. There is also a requirement to record design intent, the reasons for the design being as it is, so that future designs may be informed by what has gone before.

9 Information gathering

Two separate modes of information gathering are presented, the need for information surrounding a particular problem and the general updating of information which must be continuous on both a personal and design team basis. The general information gathering process is outlined as defining the purpose of the search, searching, locating, obtaining, rejecting the irrelevant, filing the information and highlighting for easy reference. In the definition of a PDS the information which must be consolidated starts with the design brief and includes the context of the product, matters of confidentiality, the product development and any specific company requirements. Sources of information are identified as libraries, encyclopaedias, handbooks, journals, indexes, component suppliers, standards, patents and databases.

9.1 Introduction

When starting to define a problem and develop a specification the first flood of information will occur in a random manner. It is essential, therefore, that everything is recorded, irrespective of whether or not the relevance of the information is immediately recognized. The list of questions and answers should subsequently be examined, organized under broad headings and expanded as thought appropriate. In effect, initial thoughts and information should be recorded in an organized manner.

The knowledge immediately available to an engineering designer can only represent a very small proportion of the total accumulated scientific and technical knowledge which may be employed in the solution of any particular problem. This is so even though ready access to sources of information, such as computer-based data through CD-Roms and Internet, books, standards and manufacturer's data can give the false impression that a particular designer's knowledge appears comprehensive.

If you were asked to design a large bucket could you do it? If the previous chapters have achieved one of their aims then the answer to this question should be, 'No, I need more information'. The design engineer should immediately ask questions such as what is the bucket to contain? Does it need a spout? How large will it need to be? What temperatures will the bucket be subjected to? Who are the customers to be? etc. So many factors influence the engineering design of products and processes that a systematic approach to information gathering is essential. Without such an approach, important factors may be overlooked, the design may take a long time or be less complete.

9.2 Continuous information gathering

The average designer devotes 10% of their time to searching for information.

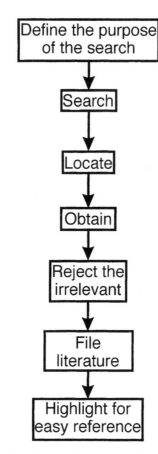

Figure 9.1 Information gathering flow chart

Searching for information is a necessary evil although regarded as the major non-productive part of a designer's working hours. The aim therefore must be to carry out the information seeking task as efficiently as is possible. Outlined in Fig. 9.1 are the essential stages in any information search. There must be a clear identification of the purpose of the search and the type of information which is required. Then the search is carried out, information located and obtained. The next stage is extremely important. A search will often identify many potential sources of information and an inevitable consequence is that at least some of this information will not prove relevant to the problem at hand. This information should be discarded. It is essential that all information is filed for easy retrieval and highlighting potentially interesting areas will greatly assist this retrieval process.

Lectures, seminars and recommended textbooks cannot provide all the information required in engineering design, particularly project work undertaken as part of an educational course, and effective use must be made of many other resources. A comprehensive review of information sources is contained in the book *Information Sources in Science and Technology* by Parker and Turley and there are many other similar texts covering the same theme.

Any professional person should ensure that they keep up to date with developments in their sphere of interest. A search for information with regard to a particular project, although an essential ingredient of any project, is no more important than this continual updating. This is particularly true for scientific and technical disciplines since these subject areas develop rapidly. The Institutions of Engineering and The Engineering Council now insist that all professional or Chartered Engineers maintain a Continuing Professional Development log, in which all this activity must be recorded. In most areas of science it is relatively easy to keep pace with developments since the field of interest tends to be narrow. Even in engineering specialists, such as stressing or materials experts, encounter little difficulty in reading round the subject. However, the problems faced by a design engineer can be so diverse that it is impossible to read all the relevant literature. Therefore, the designer must optimize reading time and be as efficient as possible.

One of the most effective ways of keeping up to date is to join a relevant institution, as a student member in the first instance, such as *The (UK) Institution of Mechanical Engineers, The American Society of Electrical Engineers* or *The (UK) Institution of Engineering Designers*. As a result of membership access will be gained to all the information sources of the institution concerned including, in most cases, a comprehensive library. Information regarding relevant forthcoming conferences, seminars, lectures and exhibitions will also be circulated. The dedicated professional will make a point of attending many of the events organized where cross-fertilization of ideas can occur, particularly during discussion sessions. Institutions also publish journals which are available to members at reduced cost along with free magazines, such as *Professional Engineer* and *Engineering Designer*. These publications invariably include all major developments taking place in a particular profession.

Regular searches, probably once a month is sufficient, should also be made in libraries for new papers and books. A mass of information will inevitably be collected and it will be necessary to sift out the irrelevant. A personal file should be kept.

The sources of information most useful in tackling a particular problem are impossible to predict. They will vary with the depth of knowledge and experience of the engineering designer concerned, with the timescale of the project and the amount of data provided in the design brief or PDS. The list of sources following should not be considered as comprehensive since there are an almost infinite number of such sources. The list is nevertheless in the order which a search should follow assuming no prior knowledge on the part of the designer.

A note should be made of all sources of information searched even if the result is that no relevant information is found. This will prevent a future repeat of the exercise. Often a zero result is as valuable as a positive result since the major problem with any search is confidence that all avenues of investigation have been followed.

9.3 Information gathering for a particular PDS

In preparation for writing a PDS, the information to be consolidated is illustrated in Fig. 9.2. The information required cannot easily be categorized but falls mainly under the following broad headings.

The design brief

Within this document it is necessary to outline the four broad categories of requirements:

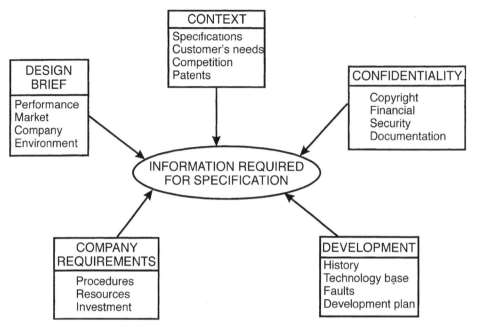

Figure 9.2 Preparation for specification writing

- *Performance requirements* such as loads, speeds and strength.
- *Market requirements* such as whether the product is targeted at mass markets, a group of specific people, individuals or a single customer.
- *Company requirements* such as the use of in-house or external resources and the use of specific processes. It is inevitable that some overlap occurs in the categories of information search. Company requirements are usually contained in the brief but often require more detailed attention for a PDS.
- *Environmental requirements* such as noise levels, emissions and recyclability.

Context

In examining the context of the product answers are sought to questions such as:

- Do national or international specifications already exist?
- Are the customers external or internal and who are they?
- Will anyone else read the PDS and do they require different information?
- Are you fully aware of the customers needs?
- Do you have up to date knowledge of competitors' products?
- Are there any patents restricting product development?

Confidentiality

The nature of the information to be included in the PDS may dictate certain levels of confidentiality. Again, answers to a series of questions such as those listed should be sought.

• Is the information required available or confidential?
• Are there copyright issues?
• Are there matters of financial confidentiality?
• Are there matters of national or organizational security that may be relevant?
• Are there company procedures with respect to the documentation of such information?

Product development

It is an advantage to have knowledge of the product history, such as previous failures or particularly successful features. Questions which should be considered include:

• In the history of the product or similar product is there knowledge of:

   faults which have occurred?
   the causes of faults?
   manufacturing or assembly difficulties?

• Is knowledge of current technological capabilities of the required level?
• Is the product development programme clearly explained and understood?

Company requirements

Again, the definition of any specific company requirements is best accomplished by attempting to answer a series of questions such as:

• Are there any company procedures for writing specifications?
• Are there adequate company resources in terms of:

   land?
   buildings?
   plant and equipment?

• Is the manpower available adequate in terms of levels of expertise and training?
• Is the level of financial investment and revenue linked to the project adequate?
• What is the expected financial or added value on this investment?
• Is a list of materials currently stocked and traditionally used available?
• Is it likely that external suppliers will be used?

9.4 Information sources

The sources of information most useful to design engineers are listed under separate headings and are mainly those available in the UK, USA and Europe. One point worth bearing in mind is that although English is the adopted language of the world-wide scientific and technical communities about 30% of all literature is produced in other languages. It may not be necessary to translate any of these as a student but professional design engineers must obtain the information somehow. Services such as *The British Industrial and Scientific International Translations Service* offers assistance in case of difficulties.

Library use

Techniques and skills have to be acquired if effective use is to be made of library facilities. The single most important tool for exploiting the resources of a library is the catalogue and a firm grasp of the conventions of the catalogue will assist an information search enormously. Books, periodicals, pamphlets and microfilms are all arranged on shelves according to classmark. Thus, in the UK, all books concerned with engineering design are shelved together under the classmark TA 174. The catalogue contains a record of every item in the library listed in three ways: alphabetical order of author's surname, an alphabetical subject index and the classified catalogue.

The task when searching for information is to identify this classmark by using either the author's surname or the subject/title index. Having identified the classmark it is not recommended to go straight to the shelves armed with the classmark. A better course of action is to look up the classmark in the classified catalogue where every item the library holds on the subject will be identified, including oversize books, pamphlets, periodicals and microfilms. Theses and dissertations are generally catalogued and stored separately.

Catalogue systems used to consist of thousands of cards containing the information and filed in drawers. Indeed many libraries still operate this type of system. However, many have now introduced a computer-based system to complement the manual system. These computer-based search systems have many more facilities and are very much quicker to use than the manual system. For example, a very useful and powerful facility is the keyword search. Using such a search all items stored in the library with the particular word or combination of words in the title can be identified.

Encyclopaedias

The *Encyclopaedia Britannica* or *Chambers Encyclopaedia* are good starting points for anyone coming new to a subject. However, it is more likely that the McGraw-Hill *Encyclopaedia of Science and Technology*, and many similar tomes, will be of use to engineers. These more specialized technical encyclopaedias tend to be written assuming that the reader has some broad technical knowledge but not necessarily about the particular subject being researched. There are now many excellent encyclopaedias available in computer-based form.

Directories

Directories provide a source of names, addresses and other information covering many subjects including the names of scientists, trades and industries. Three of the more useful directories are *Books in Print, Directory of Information Resources in U.S., Volume 1*, National Referral Center, Library of Congress and *Scientific and Technical Books in Print*. In order to identify more specialized directories *Current European Directories* and *Current British Directories* (CBD Research) should be consulted. Government publications are also an important source of information. In the UK a sales list is published annually by *Her Majesty's Stationery Office*.

To aid the identification of manufacturers and potential suppliers annual publications list companies and their products and services. These include the *Engineering Buyers Guide* (Morgan-Grampian), *The Institution of Engineering Designers Official Reference Book and Buyers Guide*, the *Thomas Register of American Manufacturers* and the *Kompass*

Register of British Industry and Commerce. Volume 1 of *Kompass* is an alphabetical product and services index. Volume 2 contains an alphabetical listing of companies containing more detailed information about each company.

Handbooks

These are generally concise reference books for day-to-day use by specialists. For engineers, handbooks such as the yearly *Kempe's* (Morgan-Grampian), the *Science Data Book*, the *Mechanical Engineers Reference Book* (Newness-Butterworth), the *Electrical Engineers Reference Book* (Butterworth), *Marks' Standard Handbook for Mechanical Engineering* (McGraw-Hill), *Standard Handbook for Electrical Engineers* (McGraw-Hill), *Electronics Engineers Handbook* (McGraw-Hill) and the *Machinery's Handbook* (Industrial Press) are just a few examples. There are many more handbooks of direct relevance to the design engineer. These vary from general design handbooks such as the *Handbook of Engineering Design* (Callum) and the *Handbook of Engineering Design using Standard Materials and Components* (Mucci) to more specialized handbooks like the *Plastics Product Design Engineering Handbook* (Levy) and the *Metals Handbook* (American Society of Metals).

Two further important source of design data are the *Engineering Sciences Data Unit* (ESDU) who publish a wide ranging series of Guides (items) which are available on subscription and the publications of *Sharing Experience in Engineering Design* (SEED). The latter include Design Procedural Guides which are intended for use during design project work.

Journals

Referring to current journals means papers which represent the most up to date thinking are seen soon after being published. Often these papers contain references to further papers which may be of interest. The papers tend to deal with specific subjects so that the information is more concentrated than it would be in a textbook. Unfortunately, it is not possible for most libraries to subscribe to all journals because of financial and space constraints. To get round this problem *Abstracting Journals* are published which include full bibliographical references helping to trace all relevant published papers. The *Monthly Catalogue of U.S. Government Publications*, Government Printing Office is worthy of mention here.

Another general reference is *Current Contents: Engineering and Technology*, Institution of Scientific Information, which is published weekly in six parts with one part covering engineering and technology. *Current Contents*, as its name suggests, consists of the contents pages of current journals along with a list of authors addresses. Most libraries have a separate periodicals holdings list which may also be consulted.

Current Research in Britain contains brief details of research in British Universities and Colleges. Volume 1 covers the physical sciences and includes engineering.

Indexes

Although these are many and varied, two broad types of index are worthy of mention here. The *Science Citation Index* (SCI) and the microfilm indexes of manufacturers' information. However, the most comprehensive index is probably the *World Index of*

Scientific Translations (World Transindex), International Translations Centre, Delft, Netherlands.

The SCI is a listing of all references (citations) found in footnotes of journals and covers approximately 15 volumes each year. It is divided into three main sections, the *Source Index*, the *Citation Index* and the *Permuterm Subject Index*. The first contains bibliographical details including author, co-author(s) and authors' addresses arranged in alphabetical order of citing (later) authors. The second is arranged alphabetically by names of cited (earlier) authors. The third is a subject index generated from the main words, not necessarily the first word, of the title.

Manufacturers' catalogues are often fruitful sources of information although some care must be exercised so as not to confuse company-specific information with general information. In the USA *MacRae's Blue Book (annual)*, a collection of manufacturers' catalogues is very comprehensive. The microfilm indexes mentioned earlier are a collection of all (subscribing) companies' catalogues on microfilm cassettes. The *Technical Index* and *VSMF Design Engineering System* microfilm systems are just two which are worthy of mention. These systems or a buyer's guide can be used to locate the catalogue of a preferred supplier, to find a possible supplier if only the product is known and to locate a supplier from a trade name. This basic service is being extended into collections of specialist files combining product and standards information. Also worthy of mention are the *Engineering Materials and Processes Information Systems*, General Electric Company.

Information gained from companies should not be restricted to that contained in catalogues. Much more can be obtained by telephone calls to sales and technical personnel, although the approach has to be right. Creating a good impression with a polite manner and by being well prepared with questions is always likely to bring better results and more information.

Standards

The following two definitions are taken from British Standard 0.

Standard

A technical specification or other document available to the public, drawn up with the cooperation and consensus or general approval of all interests affected by it, based on the consolidated results of science, technology and experience, aimed at the promotion of optimum community benefits and approved by a body recognized on the regional, national or international level.

Standardization

An activity giving solutions for repetitive application, to problems essentially in the fields of science, technology and economics, aimed at the achievement of the optimum degree of order in a given context. Generally, the activity consists of the process of formulating, issuing and implementing standards.

Standards organizations function to simplify production and distribution for manufacturers, to facilitate communication (by means of drawing standards for example), to ensure uniformity, reliability and safety for the consumer, to simplify trade across national boundaries and to promote economy of human effort by minimizing unnecessary and wasteful variety. In so doing they have become a very useful source of reliable, tested and detailed scientific and technical data. In many areas of engineering rigorous standards must be adhered to. For example, for what are primarily safety reasons, the design of pressure vessels is completely governed by specification.

The three most used standards are *ANSI* (American National Standards Institute), *ISO* (International Standards Organization) and *BSI* (British Standards Institution). As well as these most countries have their own standards which are of particular relevance to potential exporters. The three standards organizations mentioned publish a yearly catalogue with a comprehensive subject index which should be consulted in the first instance. The index leads to the main section which contains brief descriptions of the contents of each individual standard. Each yearly catalogue includes much more information. For example, the *BSI Catalogue* contains a list of libraries holding complete sets of standards.

Examples of American catalogues include the *Annual Catalogue*, American National Standards Institute, the *Annual book of ASTM Standards*, American Society for Testing and Materials, the *Index to Specifications and Standards*, United States Department of Defense and the *Index to Federal Specifications and Standards*, United States General Services Administration. European and international standards include the Deutsches Institut fur Normung (*DIN*), the *ISO Catalogue*, International Standards Organization and the *Catalogue of IEC Publications*, International Electrotechnical Commission.

In Chapter 2 covering PDS definition, operation requirements and particularly safety requirements were discussed. As an example of the sort of useful information contained in standards Fig. 2.5 was included and shows an electrical test finger.

Although not particularly relevant to a search for information outside the industrial environment the reader should be aware that most companies complement national and international standards with their own. National and international standards are inevitably wide in scope and companies often wish to use only part of the standard. For example a company may wish to rationalize the standards covering the range of drilled hole sizes, thus reducing the variety of drill bits stored. The designers would then be restricted to using only preferred sizes, unless there are exceptional circumstances.

Patents

New ideas can earn millions for inventors if they are protected by patents. The inventor of the ring-pull opener for cans still receives a small payment every time a can is sold! However, this section is not about how to protect a new idea, rather it is about the use of patents as a valuable information source. To assess the value of searching patent literature as a means of satisfying your information needs it is necessary to be aware of the attributes and shortcomings of the information which can be gained.

Every patent must add to the state of the art in the particular field. The description or specification discloses an invention in sufficient detail for another person to be able to repeat the invention and includes the reasoning behind the invention. The accompanying search report or citations may give references which lead to further relevant information. As an example of a patent, GB patent no. 2171746A showing a novel type of ladder stabilizer is included here as Fig. 9.3.

Reading a patent can be difficult since a patent is a legal document which must stand up to challenge in the courts. Also, standard industrial technology is rarely presented and the document concentrates solely on the novelty contained in the invention. Any worked examples assume expert knowledge on the part of the searcher. Nevertheless, a sound

(12) **UK Patent Application** (19) **GB** (11) **2 171 746 A**

(43) Application published 3 Sep 1986

(21) Application No **8604704**

(22) Date of filing **26 Feb 1986**

(30) Priority data
(31) **8505289** (32) **1 Mar 1985** (33) **GB**

(71) Applicant
Derek Short,
20 Tebay Close, Overfields, Middlesbrough, Cleveland

(72) Inventor
Derek Short

(74) Agent and/or Address for Service
Urquhart-Dykes & Lord,
New Exchange Buildings, Queen's Square,
Middlesbrough TS2 1AB

(51) INT CL⁴
E06C 7/46

(52) Domestic classification (Edition H):
E1S LW2

(56) Documents cited
GB 0656438

(58) Field of search
E1S
Selected US specifications from IPC sub-class E06C

(54) **Ladder stabiliser**

(57) A ladder stabiliser for stabilising the lower end of a ladder in use comprises at least one linearly extensible elongated member (5,6), a pair of ladder sockets (10,11) spaced apart transverse to the length of said member and fixed or fixable relative to the length of said member, each ladder socket being adapted to receive the lower end of one side of a ladder, and ground-engaging means (20,21) on the elongated member. In an illustrated preferred form, the stabiliser comprises two such elongated members, spaced apart by a cross-member and each having a ladder socket.

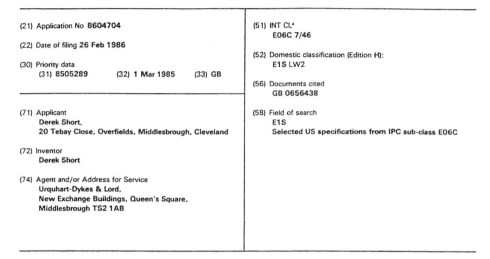
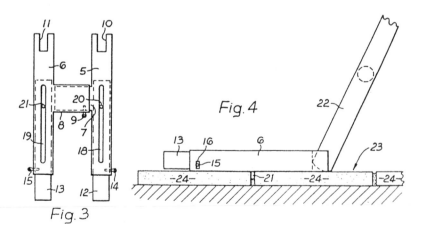

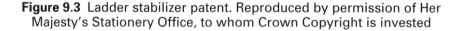
Figure 9.3 Ladder stabilizer patent. Reproduced by permission of Her Majesty's Stationery Office, to whom Crown Copyright is invested

1/2

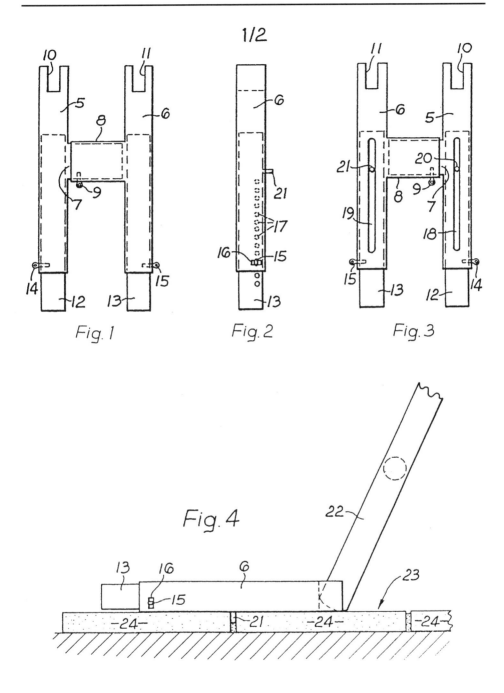

Fig. 1   Fig. 2   Fig. 3

Fig. 4

Figure 9.3 continued

2/2

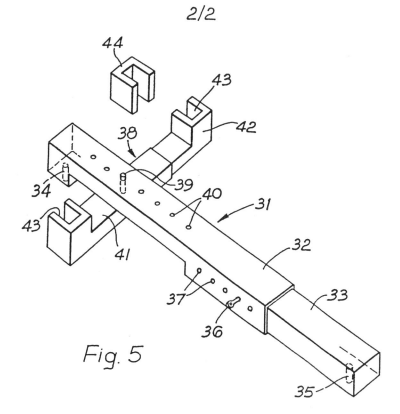

Fig. 5

Figure 9.3 continued

flagstones.

Other forms of ground-engaging means are possible, though less preferred. For example, such means may be in the form of heavy
5 serrations or saw-teeth or similar projections.

Desirably, the material of which the ladder stabiliser is made should combine the necessary strength with lightness. To that end, I particularly prefer to make it in aluminium.
10 Thus tubular aluminium is especially suitable, in circular or square cross-section.

My invention will now be further described with reference to the accompanying drawings, in which:
15   Fig. 1 is a plan view of one form of ladder stabiliser according to my invention;

Fig. 2 is a side elevation from the right-hand side of Fig. 1;

Fig. 3 is a view from the underside of the
20 ladder stabiliser of Figs. 1 and 2;

Fig. 4 shows the ladder stabiliser of Figs. 1 to 3 in use; and

Fig. 5 is a perspective view of a second form of ladder stabiliser according to my in-
25 vention.

The ladder stabiliser shown in Figs. 1 to 4 of the drawings comprises two tubular, aluminium elongated members 5, 6 disposed in generally parallel alignment and spaced apart
30 by a telescopic cross-bar. The cross-bar consists of tubular lateral projections 7, 8 from the members 5, 6. The tubular projection 7 is a close sliding fit within the tubular projection 8 and may be secured relative thereto by a
35 split pin 9, which engages in one of a series of holes in the projection 7. Telescopic adjust-ment of the cross-bar permits the spacing apart of the elongated members 5, 6 to be adapted to the width of the ladder to be sta-
40 bilised. The split pin 9 may be omitted if de-sired, since the forces transmitted to the sta-biliser by the ladder in use are at right angles to the cross-bar and do not therefore tend to affect the spacing of the members 5, 6.
45   The members 5, 6 have terminal open slots 10, 11, each wide enough to receive the lower end of one side of a ladder.

Within the elongated members 5, 6 and a close sliding fit therein are two tubular alumi-
50 nium extensions 12, 13. Each of these exten-sions may be locked in one of several ex-tended positions relative to the respective elongated member by a split pin 14 or 15. As more clearly seen in Fig. 2, the split pin 15
55 extends through a cut-out 16 in the member 6 and is located in a selected one of a series of holes 17 in the extension 13. The extension 12 is secured in a similar manner.

On the underside of each of the elongated
60 members 5, 6 is an elongated slot 18 or 19, through which projects a pin 20 or 21 mounted on the extension 12 or 13 respec-tively. Each pin is exchangeable for at least one alternative pin of greater length.
65   Fig. 4 shows the stabiliser of Figs. 1 to 3 in

use to support a ladder 22 upon a ground surface 23 consisting of flag-stones 24. In practice, a convenient way of working is first to stand the ladder 22 in its desired position
70 of use, then to place the stabiliser in position with the slots 10, 11 enclosing the feet of the ladder and to adjust the extensions 12, 13 until the pins 20, 21 can be located in one or two gaps between flagstones. By means of
75 the split pins 14, 15 the extensions may then be secured in position.

The alternative form of ladder stabiliser illus-trated in Fig. 5 of the drawings comprises a single elongated member 31, in the form of
80 an outer member 32 of generally rectangular cross-section and an inner member 33 of matching cross-section slidable therein so as to project by a selected distance therefrom. Downwardly-extending pins 34 and 35 project
85 from the lower face of the members 32 and 33 respectively so as to engage the ground and retain the stabiliser in place. The distance apart of the pins 34 and 35 is altered by relative telescopic adjustment of the members
90 32 and 33. The latter members are relatively secured, after adjustment, by a split pin 36 extending through a selected one of a series of apertures 37 in the outer member 32 into a socket in the inner member 33.
95   A cross-member 38 is secured in position transverse to the member 31 by means of a pivot pin 39 standing upwardly from the cross-member 38 through a selected one of a series of apertures 40 along a length of the
100 member 31. The cross-member 38 comprises a fixed outer sleeve 41 and an inner member 42 linearly slidable therein. The sleeve 41 and member 42 each carries a socket 43 to re-ceive the lower end of a ladder side (not
105 shown). Added security may be afforded, if desired, by means of attachments 44 (only one of which is shown), which may be bolted across the sockets 43 to fully enclose the respective ladder feet.
110   To use this second form of ladder stabiliser, the ladder is first placed in the desired posi-tion and the cross-member 38 is telescopically adjusted until the sockets are so spaced apart as to engage the lower ends of the ladder
115 sides. The member 31 is now placed over the cross-member 38 in such a position that the pin 34 can engage the ground at a convenient point and the pivot pin 39 projects through an adjacent aperture 40. The position of the inner
120 member 33 within the member 32 is adjusted linearly until the pin 35 can engage the ground at a desired point; the members are then se-cured by the split pin 36.

It should be noted that further adjustability
125 is afforded by turning the cross-member through 180° so as to face the opposite di-rection.

The stabiliser according to my invention, for example the embodiments illustrated in the
130 drawings, affords convenient and ready sup-

Figure 9.3 continued

port for a ladder without the need of a second person to support the ladder while it is in use.

5 CLAIMS

1. A ladder stabiliser comprising at least one linearly extensible elongated member, a pair of ladder sockets spaced apart transverse to the length of said elongated member and
10 fixed or fixable relative to the length of said elongated member, each said ladder socket being adapted to receive the lower end of one side of a ladder, and ground-engaging means on said elongated member.

15 2. A ladder stabiliser as claimed in claim 1, wherein the ground-engaging means is adjustable relative to the ladder sockets in a direction parallel to the length of the elongated member.

20 3. A ladder stabiliser as claimed in claim 2, comprising a pair of said extensible elongated members, each said member having a said ladder socket and said groundengaging means being on said elongated members.

25 4. A ladder stabiliser as claimed in claim 2, comprising a single linearly extensible elongated member and a cross-member having both said ladder sockets thereon, the ground-engaging means being on the elongated mem-
30 ber.

5. A ladder stabiliser as claimed in claim 3, wherein the extensible elongated members are spaced apart by a cross-member.

6. A ladder stabiliser as claimed in claim 4
35 or claim 5, wherein the cross-member is extensible to permit relative adjustment of said ladder sockets.

7. A ladder stabiliser as claimed in any of the preceding claims, wherein the sockets are
40 open-ended slots.

8. A ladder stabiliser as claimed in any of the preceding claims, wherein the extensible elongated members each comprises a hollow outer member and an inner member telescopi-
45 cally disposed therein.

9. A ladder stabiliser as claimed in claim 8, wherein the relatively movable parts of each said extensible elongated member are relatively fixable in at least one extended position.

50 10. A ladder stabiliser as claimed in any of the preceding claims, wherein the ground-engaging means is at least one pin projecting from the elongated member.

11. A ladder stabiliser as claimed in any of
55 claims 1 to 9, wherein the ground-engaging means comprise serrations or saw-teeth.

12. A ladder stabiliser as claimed in any of the preceding claims, wherein the ground-engaging means is located upon an extensible
60 part of the elongated member.

13. A ladder stabiliser substantially as hereinbefore described with reference to, and as illustrated in, Figs. 1 to 4 of the accompanying drawings.
65 14. A ladder stabiliser substantially as here-

inbefore described with reference to, and as illustrated in, Fig. 5 of the accompanying drawings.

Printed in the United Kingdom for
Her Majesty's Stationery Office, Dd 8818935, 1986, 4235.
Published at The Patent Office, 25 Southampton Buildings,
London, WC2A 1AY, from which copies may be obtained.

Figure 9.3 continued

understanding of engineering principles is sufficient background knowledge in most cases. When completely new products are invented, such as the Hovercraft, the whole machine is described in great detail. This is relatively unusual and most patents refer only to components of a machine affected by the invention. Thus, when using keywords for searching, component names should be used as well as those words which define the whole machine.

The layout and content of a British patent are best explained with reference to the ladder stabilizer. Other countries patents vary slightly in the way the information is organized, but the content is substantially the same. The major sections are:

(11)   The patent number which is a unique national number.

(51, 52) Each patent specification is given an international classification (IPC) and one (or more) national classification. This enables the subject of a patent to be identified by use of the Classification Key and the Catchwords Index.

(43)   Each patent specification gives various dates, the last being the date the patent was obtained.

(71)   The name of the applicant which may be found by searching the Applicants Name Index.

(56)   List of documents cited.

(54)   The title.

(57)   An abstract, an explanation of what is already known (prior art), the object of the invention, detailed description with drawings and finally an itemized list of claims.

Figures released by patent offices claim that approximately 80% of information contained in patents is not available through any other source. Also, of the four million patents issued to date, 85% are no longer restricted for use. Once a patent has expired or been allowed to lapse anyone may make use of the information disclosed. Patents are comprehensively classified and indexed and in the supplementary indexes of the SCI a *Patent Citation Index* lists patents by their serial number. Year of issue, name of patentee and country of origin are also quoted. The *Official Journal (Patents)* is published weekly in the UK. In America the following sources are useful. The *Official Gazette (weekly)*, United States Patent Office, the *Annual Index*, United States Patent Office and the *Public Patent Search Facilities*, Patent and Trademark Office, Arlington, Virginia.

The computer database search and advisory service gives access to all patents, past and present. When making a patent search it is essential that the technical subject to be searched is expressed as precisely as possible. This is a very important research tool and has the potential for saving much time and money. Incredible as it may seem, there are plenty of examples of companies working for long periods on solving a particular problem only to find on conducting a patent search that a lapsed patent solved the problem for them. There have even been cases of a company discovering that they themselves hold the relevant patent and had completely forgotten!

Databases

These are still thought of as computerized equivalents of abstracting or indexing journals, directories and handbooks. However, databases have begun to evolve away from their hard-copy equivalents in that they often contain extra information to enhance on-line searching and already have no hard-copy alternative.

One form of database is the videotext which is usually taken to mean the information available through a domestic television receiver. At present these services are not a useful source of technical information, but the services may be expanded to offer transmissions for home computers, including business and educational software. Viewdata is the name used for services available via the telephone. Again the amount of useful technical data is limited at present although these services can often be used as the gateway to other services permitting full interactive communication.

The British Standards Institute have introduced an on-line database called *BSI STANDARDLINE*. There are many other databases too numerous to mention. However, those which are potentially most useful to engineers are *NTIS*, Government Reports Announcements and Index, updated bi-monthly, National Technical Information Service, United States Department of Commerce; *Metadex*, ASM & The Metals Society (London); *Compendex*, Engineering Index; *EKOL*, European Kompass On-Line; *KWIC* Index of International Standards, covering ISO, IEC and most national standards; and *ISONET*, bibliographical descriptions of standards and regulations organized by ISO.

Finally, of course, as the Internet (or World Wide Web) becomes more accessible and search engines become increasingly powerful then this will increasingly provide much of the information required. Already, many supplier companies have their own sites through which components can be selected and ordered. With the bookmarking facilities available then information sources which need to be accessed regularly can also be accessed rapidly. It is inevitable that the Internet will grow in importance as a major source of information. However, at present there are no guarantees that the information obtained is correct!

As an example of the range of information consider just one site. The address is http://www.globalspec.com/ and what was obtained at the time of searching is shown in Fig. 9.4, although obviously with links to many other sites and in glorious technicolour.

If the user selects *Compression Springs*, for example, then they are presented with an on-line form into which they type the known characteristics of the required spring. Guidance is given regarding many aspects of a spring, including length to diameter ratio and materials available. The system searches all spring manufacturers with data in the Internet.

9.5 Principles

Information gathering principles

Questioning   In the early stages of information gathering the answers to many questions are sought. Care must be taken to verify the answers obtained.

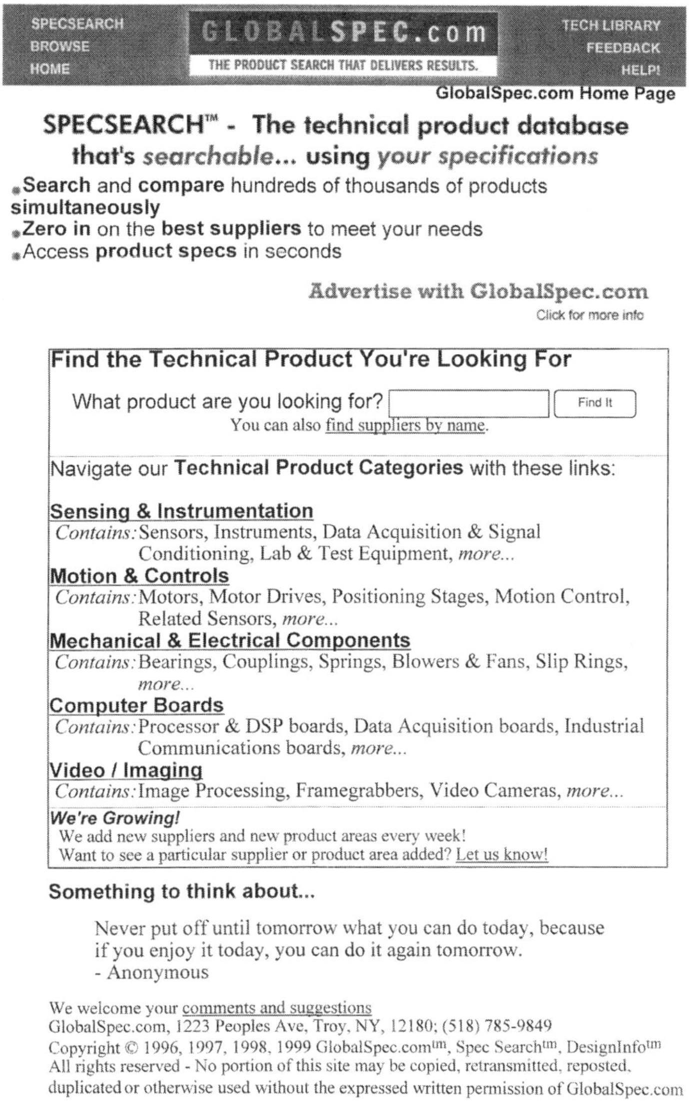
SPECSEARCH
BROWSE
HOME
GLOBALSPEC.com
THE PRODUCT SEARCH THAT DELIVERS RESULTS.
TECH LIBRARY
FEEDBACK
HELP!
GlobalSpec.com Home Page
SPECSEARCH™ - The technical product database that's searchable... using your specifications
•Search and compare hundreds of thousands of products simultaneously
•Zero in on the best suppliers to meet your needs
•Access product specs in seconds
Advertise with GlobalSpec.com
Click for more info
Find the Technical Product You're Looking For
What product are you looking for?
Find It
You can also find suppliers by name.
Navigate our Technical Product Categories with these links:
Sensing & Instrumentation
Contains: Sensors, Instruments, Data Acquisition & Signal Conditioning, Lab & Test Equipment, more...
Motion & Controls
Contains: Motors, Motor Drives, Positioning Stages, Motion Control, Related Sensors, more...
Mechanical & Electrical Components
Contains: Bearings, Couplings, Springs, Blowers & Fans, Slip Rings, more...
Computer Boards
Contains: Processor & DSP boards, Data Acquisition boards, Industrial Communications boards, more...
Video / Imaging
Contains: Image Processing, Framegrabbers, Video Cameras, more...
We're Growing!
We add new suppliers and new product areas every week!
Want to see a particular supplier or product area added? Let us know!
Something to think about...
Never put off until tomorrow what you can do today, because if you enjoy it today, you can do it again tomorrow.
- Anonymous
We welcome your comments and suggestions
GlobalSpec.com, 1223 Peoples Ave, Troy, NY, 12180; (518) 785-9849
Copyright © 1996, 1997, 1998, 1999 GlobalSpec.com™, Spec Search™, DesignInfo™
All rights reserved - No portion of this site may be copied, retransmitted, reposted, duplicated or otherwise used without the expressed written permission of GlobalSpec.com

Figure 9.4 GlobalSpec.com home page. Reproduced with permission of GlobalSpec.com

Sifting There is too much information available. It is necessary to sort the important from the unimportant.

Highlighting Not all information important enough for filing will be of the same degree of usefulness. For ease of future reference those sources of information which will need to accessed frequently should be highlighted.

Updating It is the duty of all professional engineers to ensure that the information they have is up-to-date.

Confidence If at all possible information should be obtained from at least two independent sources in order to ensure the accuracy of that information.

9.6 Exercises

1. Using all resources at your disposal search for and list relevant information sources for a selection of the eight exercises introduced at the end of Chapter 2 or the five new exercises at the end of Chapter 3. No detailed information is required, only a listing of information sources. Your approach will differ depending upon the level of prior knowledge and understanding you have of the chosen subject. The first step should be a clear statement of the aims of the search. It is suggested that you start the search with encyclopaedias, handbooks and directories, then look for relevant books and use abstracting journals, next search for the names of organizations with interest in the subject and finally list relevant standards and codes of practice, patents and use buyer's guides and the Science Citation Index. Any other sources you think relevant should also be searched.

10 Presentation techniques

Drawing morphology, types of engineering design drawings and graphical presentation are covered. The aim is to enable you to present a coherent design report with drawings which will satisfy a 'customer', even if that 'customer' is a design office manager or technical director. The reasons for keeping clear records have been explained in earlier chapters: what constitutes a thorough design project report is presented here. The types of drawings and reports are illustrated by two further case studies.

10.1 Introduction

Every stage of design requires some form of drawing to support it. The type of drawing depends upon the purpose of the drawing and the type of information it contains. Clearly, it would not be an efficient use of precious resources to produce final drawings of a product, if the designer has not yet formulated likely concepts during an earlier creative stage. During the creative stage rapid and imaginative drawing and sketching techniques are fundamental to a successful outcome. It is worth considering the morphology of drawing, outlined in Fig. 10.1, to identify the purpose and techniques appropriate to a given purpose.

Modelled property	Receiver	Code	Drawing technique	Drawing types
Function	The designer	Co-ordinates	Free hand sketching	Block diagram
Structure	Another designer	Symbols	Draughting machine	Graph
Form	Technical draughtsman	Rules	Use of templates	Concept sketch
Material	Workshop staff	Projection	2D CAD	Drawing of principle
Dimension	Production planner		Surface modeller	Scheme
Surface	Manager		Solid modeller	Detail drawing
	Customer			Assembly drawing
	Professional group			Exploded drawing
	Public authority			Patent drawing

Figure 10.1 Drawing morphology

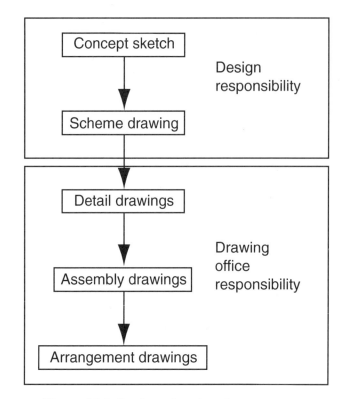

Figure 10.2 Engineering drawing sequence

The four most important characteristics of drawing: modelled property, receiver, code and drawing technique, are identified which when combined indicate the most appropriate type of drawing. As illustrated, to communicate with fellow designers sketched concepts or drawings of principle are usually best.

Figure 10.2 illustrates where design responsibility ends, except for the checking of drawing office produced drawings, and draughtsman responsibility begins. The design team uses sketches to communicate ideas during the early stages of the project. Some of these sketches are drawn to a very high standard, particularly those to be viewed by 'customers'. However, most of the sketches are for discussion with other engineers and need not be produced to such exacting standards.

The design team also uses proving drawings which, for example, show the extremities of motion. The output from the design team is usually in the form of calculations and scheme drawings giving all the information required for production drawings and instructions to be developed. Some further documentation is often required since the 'design intent' also needs to be communicated.

The scheme drawing and all the drawings which follow are governed by the conventions given in BS 308.

10.2 Concept sketches

In the very early stages of a design project, following full definition of the problem, it is usual for ideas and concepts to be communicated by means of simple sketches such as those for the seat suspension mechanism in Fig. 3.12. They are generally not drawn to scale and are made up of conventional signs and symbols to indicate component parts. Their purpose is to assist in the understanding of how the proposed system would work. These sketches can be developed further into neat and clear sketches, such as those illustrated in Fig. 4.3 for fixing a gear to a shaft. Such sketches represent the first stage of the long design process and it is essential that engineering designers develop rapid sketching skills. During the embodiment stage of the project the requirement for more detailed representations of design ideas grows and good quality sketches are required for communication to 'customers' who may not be engineers. The sketching skills developed by design engineers should encompass 3D, at least isometric unencumbered by perspective, for this purpose.

It is unusual in these early stages for the representation and development of ideas to be assisted by the use of a computer. Indeed it is generally accepted that the rigour required to operate and 'sketch' with a computer runs counter to the creative process and actually stifles it. However, if the design project is one which involves the development and improvement of an existing product then consideration of alternative designs can usefully be supported by superimposing new proposals over the existing design.

As an example consider the concept drawings for the improvement of the cable tray design in Fig. 10.3 (a). Such trays are supported by means of hangers in false ceilings and can carry quite heavy cables. The existing design is suspected of being manufactured from steel which is too thick. However, in some isolated instances the sides collapse inwards causing failure. Alternative concepts have to take into account the method of manufacture which is that of rolling from strip then cutting to length. The concepts illustrated in Fig. 10.3 (b) and (c) involve change of shape and therefore cross-sectional area of cables which can be carried. However, in both cases stiffness (indicated by finite element analysis) is improved.

The idea in Fig. 10.3 (d) is to make the tray into a box shape and therefore stiffer by 'clipping' in a top section. However, the point being illustrated is that the existing tray can be used as a comparison with new designs by drawing directly into a CAD package.

One of the first drawing sheets during the creative phase of the design process is a sheet which expresses graphically and in notes the design brief. Often it is worth drawing the known geometry to scale at this stage since without scale concepts can be difficult to

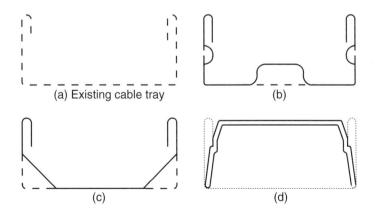

Figure 10.3 Concepts for cable tray

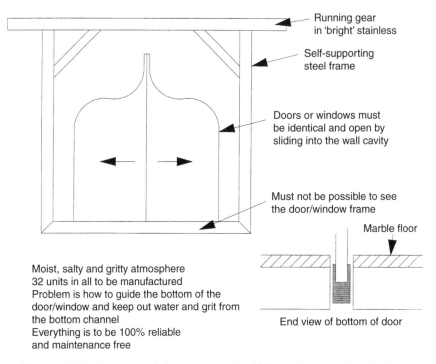

Running gear
in 'bright' stainless

Self-supporting
steel frame

Doors or windows must
be identical and open by
sliding into the wall cavity

Must not be possible to see
the door/window frame

Marble floor

Moist, salty and gritty atmosphere
32 units in all to be manufactured
Problem is how to guide the bottom of the
door/window and keep out water and grit from
the bottom channel
Everything is to be 100% reliable
and maintenance free

End view of bottom of door

Figure 10.4 Design brief – automatic sliding doors and windows

evaluate objectively. As an example, consider a project to provide automatic sliding doors and windows for a very expensive individually designed house which is directly on a beach by the sea.

The architect specified most of the building details, the shape and style of the windows and the size of the cavity. These then formed part of the design brief and the specification. The geometry of a door and some of the criteria on which the design was to be based are illustrated in Fig. 10.4. The main view indicates the preliminary opinion that a frame would be required, hidden within the cavity wall, upon which the doors would hang and 'run'. The enlarged view of the bottom of the door and the marble floor indicates one of the major design problems: the fact that the 'customer' did not want a step in marble, so no frame could be visible on the floor.

Alternative concepts and thought for the floor area are presented in Fig. 10.5. Clearly the sketches have been tidied up for presentation here but the principle that simple, incomplete sketches are sufficient at this stage in the design process is amply illustrated.

10.3 Scheme drawing

The main type of drawing used by a designer is a scheme drawing. This also forms the bulk of the information transfer to those dealing with the production details. These are generally two-dimensional drawings used in the design process to develop the ideas. They give all the essential information for a design to develop further. They show the extreme positions and arcs through which moving parts travel. Generally they are not complete drawings in

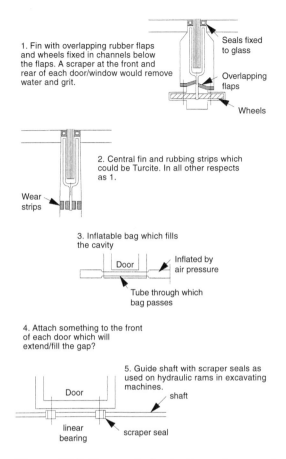

1. Fin with overlapping rubber flaps and wheels fixed in channels below the flaps. A scraper at the front and rear of each door/window would remove water and grit.

Seals fixed to glass

Overlapping flaps

Wheels

2. Central fin and rubbing strips which could be Turcite. In all other respects as 1.

Wear strips

3. Inflatable bag which fills the cavity

Door

Inflated by air pressure

Tube through which bag passes

4. Attach something to the front of each door which will extend/fill the gap?

5. Guide shaft with scraper seals as used on hydraulic rams in excavating machines.

Door

shaft

linear bearing

scraper seal

Figure 10.5 Concepts for bottom door seal

the sense that, for example, they may show the position and size of a spring rather than the drawing of the spring itself.

Figure 10.6 shows part of a scheme drawing in which a handle is to be moved through a pre-determined arc and has three discrete stop positions. The figure shows only the front elevation and end and/or side elevations would be developed at the same time. In drawing such views, which arc drawn strictly to scale, the designer is trying to ensure that there are no clashes between the handle and other components in an assembly and that the required motion is provided.

To illustrate how part scheme drawings are used which eventually lead to full production drawings consider the design of a new tilt tray sorter. A whole series of these carriages with tilting trays are employed, amongst many other applications, in airport baggage handling systems. In essence, they travel around a track with a bag on top and when they reach the correct chute for a particular aircraft, tilt and discharge the bag. A critical part of the design is therefore ensuring that the angle of tip is sufficient to dislodge all types and weights of bag. The front elevation of the part scheme drawing developed for this purpose is given in Fig. 10.7.

The scheme drawing was used for many purposes, not least to try to optimize the tilting mechanism, the supporting structure and the positioning and angles of the wheels. As can

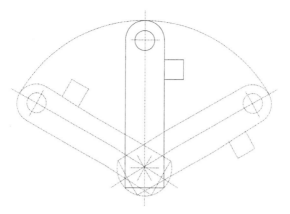

Figure 10.6 Scheme drawing showing motion

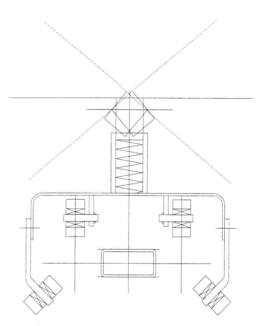

Figure 10.7 Tilt tray tipping mechanism

be seen from the assembly front elevation of Fig. 10.8, two horizontal and two vertical were finally chosen.

10.4 Design report

Along with concept sketches and a scheme drawing a full design report should contain the following sections:

Title A title sheet will normally start the report and should include the title of the design

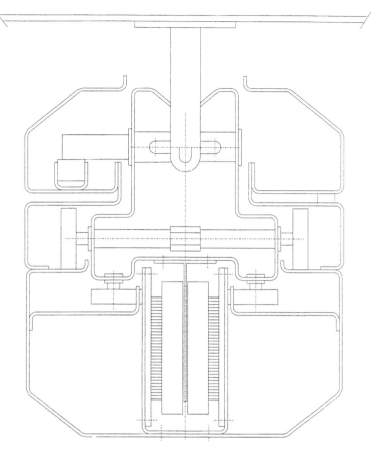

Figure 10.8 Assembly front elevation for tilt tray sorter

project, the names of those involved in the project and the date. Any information provided at the start of the project should be included after the title sheet.

Summary A single sheet describing the project, indicating how the project progressed and any difficulties encountered and giving the final outcome.

Contents A listing of the written sections, figures and drawings.

Project management Planning and control charts, such as a Gantt chart, along with minutes of all meetings and design audits. Often this is best presented as an appendix.

Information sources Information is used throughout a design project and many of the sources of information are simply referenced. However, it is often necessary to include some detailed information, normally in an appendix which is cross-referenced to the main body of the report.

Product Design Specification This is a fluid document which is developed throughout the progression of a design project. However, a definitive version, agreed with the 'customer' should be presented in the report. This contains a full definition of the function(s) and constraints.

Concepts All generated concepts presented to the same level of detail with both sketches and explanatory notes.

Concept selection Ranking and weighting of the specified constraints followed by detailed evaluation of each concept. This section should include a discussion of the main deciding factors.

Embodiment More detailed sketches and development of the selected concept. Form design, shape, ergonomics and aesthetics sections as appropriate.

Modelling (analysis) Includes all relevant calculations with illustrations as appropriate. All assumptions made indicating the limits of feasibility must be stated.

Detail design The output is sufficient detail, in the form of scheme drawings and notes, to enable production drawings to be completed.

Conclusions The final summary of the project as a whole. Give a statement of what has been accomplished and recommendations for future work.

10.5 Principles

Presentation techniques principles

Graphical representation Identify and use appropriate techniques and drawing types for target audience. Every designer must develop skills in sketching and accurate drawing.

Textual information Clarity and simplicity of style is the key to conveying information succinctly and accurately.

Index